Mechanistic Home Range Analysis

MONOGRAPHS IN POPULATION BIOLOGY
EDITED BY SIMON A. LEVIN AND HENRY S. HORN

A complete series list follows the index.

Mechanistic Home Range Analysis

PAUL R. MOORCROFT

AND

MARK A. LEWIS

PRINCETON UNIVERSITY PRESS
Princeton and Oxford

ISBN-13: 978-0-691-00927-8 (alk. paper)

ISBN-10: 0-691-00927-9

ISBN-13 (pbk.): 978-0-691-00928-5 (pbk. alk. paper)

ISBN-10 (pbk.): 0-691-00928-7

Library of Congress Control Number: 2006925777

British Library Cataloging-in-Publication Data is available

This book has been composed in Times Roman
Printed on acid-free paper. ∞

pup.princeton.edu

Printed in the United States of America

10 9 8 7 6 5 4 3 2 1

Contents

Appendixes

Preface

Over the same period that radio telemetry has provided a wealth of information on the ways in which animals move, interact, and utilize their environment, a separate mathematical literature on mechanistic models of animal movement has developed based on the analysis of correlated random walks. These two areas of research have remained largely independent of each other for two reasons. First, the mathematics involved in models of animal movement typically employs partial differential equations, which lie outside the mathematical training of most animal ecologists. Second, the general analytic intractability of such models in two space dimensions has meant that most theoretical analyses of movement equations have been restricted to a single space dimension. While convenient mathematically, the inevitable loss of realism that accompanies this simplification is off-putting to many animal ecologists. As a result, the analysis of empirical home range data has relied almost exclusively on purely descriptive statistical methods.

Our goal in writing this book is to demonstrate how mechanistic home range models, derived from underlying correlated random walks of individual movement behavior, can unify theoretical and empirical home range analysis and, in doing so, provide an ideal framework for developing a predictive, rather than merely descriptive, theory of animal home range patterns. Here, we use this approach to develop and test a series of mechanistic home range models for patterns of space use in carnivores. However, similar approaches could be used to investigate home range patterns in a variety of different animal groups.

There have been two important developments that make such a synthesis possible. First, through the pioneering work of Akira Okubo, Simon Levin, Peter Karieva, Peter Turchin, and others, there is greater awareness within animal ecology of how correlated random walks can be used as a basis for mechanistic models of animal movement. Second, modern computing power and numerical methods allow for the solving of the complex, formerly intractable systems of partial differential equations that typically result from correlated random walk models for animal movement in two dimensions. As we show through the course of this book, it is now possible to explore biologically realistic models of home range behavior that can be directly compared to empirical observations of animal movement paths and patterns of space use via numerical maximum likelihood methods. Simplified implementations of these same models can then be analyzed mathematically to elucidate general principles and conclusions regarding the patterns of space use seen in different populations.

This book is directed at both animal ecologists and mathematicians interested in understanding animal home range patterns. With this in mind, we have endeavored to make our presentation accessible to people with minimal mathematical training. However, readers wishing to follow the mathematical derivations will require a basic working knowledge of partial differential equations.

P.R.M. wishes to especially thank his thesis advisers Steve Pacala and Dan Rubenstein for convincing him that the bridge between theoretical and empirical ecology could and should be built, and giving him the intellectual freedom and wherewithal to pursue his interest in carnivore ecology. M.A.L. wishes to thank Jim Murray for his mentorship and ideas that were crucial in developing the early stages of this research.

We thank Jenny Sheldon, who read the entire manuscript and provided many constructive suggestions for improvement, and John Varley of the National Park Service for his continuing support of carnivore research in Yellowstone. We also thank Simon Levin, Jim Murray, and our respective colleagues at Harvard University and the University of Alberta for their support and encouragement for writing this book.

We are indebted to Robert Bechtel, Jeremy Cesarec, Chris Preheim, and Kym Schreiner for reading and commenting on drafts of the manuscript and to Dan Lipsitt for his assistance with some of the computer simulations. We are both grateful to Sam Elworthy, Hanne Winarsky, and Alycia Somers at Princeton University Press for their support and assistance in bringing this book to publication and to Jodi Beder editor for help with copyediting.

The advice and support we received outside of the workplace deserves a special note of gratitude. P.R.M. would like to thank Wendy Chun for her editorial skills, which improved the clarity of his writing, and for her loving support during the many hours he spent working on this book. M.A.L. would like to thank Allison for her encouragement and help in finding the essential balance between writing the book and life with children. This work was supported in part by U.S. National Science Foundation Grants, Canada Natural Sciences and Engineering Research Council of Canada grants, and a Canada Research Chair to M.A.L. and a Princeton University Fellowship to P.R.M.

We dedicate this book to our colleague and dear friend Bob Crabtree for his unique insights and immeasurable help to us both over the years since P.R.M. first arrived at the trailer in Cook City to begin studying carnivores in

Yellowstone with Bob. We especially appreciate Bob's generosity in allowing us to analyze the coyote telemetry datasets collected in Yellowstone and Hanford that have been invaluable in demonstrating that mechanistic home range models can be used to analyze real carnivore home range patterns.

P.R.M., Harvard University
M.A.L., University of Alberta

Mechanistic Home Range Analysis

CHAPTER ONE

Introduction

The advent of radio telemetry in the late 1950s revolutionized the study of animal movement, enabling the systematic measurement of animal movement patterns (Cochran and Lord 1963). Following its introduction, telemetry rapidly became a mainstay in wildlife studies and now is routinely used to track the movements of a variety of animals, including ungulates, rodents, primates, and carnivores (Macdonald et al. 1980; Millspaugh and Marzluff 2001). Telemetry has also been successfully used to study the movements of birds, reptiles, amphibians, fish, and even insects (Priede and Swift 1993). The recent development of global positioning system (GPS)–based telemetry is further enhancing its scope, allowing ecologists and wildlife biologists to accurately track animal movements over any distance, under all weather conditions, and in any terrain (Rodgers et al. 1996; Girard et al. 2002).

As in other mammalian groups, telemetry studies have documented a diverse array of patterns of space use among carnivores (Macdonald et al. 1980). For example, figure 1.1 shows estimates of home range size for three northern hemisphere canids: red foxes (*Vulpes vulpes*), coyotes (*Canis latrans*), and gray wolves (*Canis lupus*). The home range sizes of these three species alone vary over three orders of magnitude, from a few square kilometers to over a thousand square kilometers, reflecting both intra- and interspecific differences in home range size: red fox home ranges vary between 2 and 16 km^2, overlapping the distribution of coyote range sizes, which span a 30-fold range between 2.5 and 70 km^2. Wolf territories are larger still, varying between 80 and 1800 km^2.

While the between-species differences in home range size seen in figure 1.1 can be accounted for by differences in body size (Gittleman and Harvey 1982; Gittleman 1985; Kelt and Vuren 2001), the substantial degree of intraspecific variation in home range size exhibited by all three of the species is testimony to the ability of carnivores to adjust their patterns of space use in response to the local environmental and social conditions they experience (Sheldon 1992). In addition to varying in size, carnivore home ranges also vary in degree of exclusivity, from loosely defended overlapping home ranges to non-overlapping territories that may be defended by individuals, pairs, or groups (Macdonald and Moehlman 1973; MacDonald 1983; Moehlman 1989). As in other groups

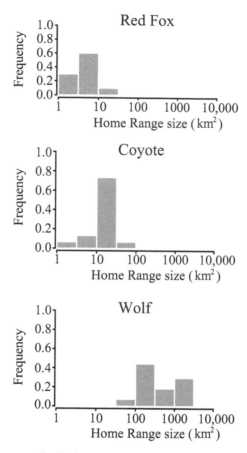

FIGURE 1.1. Frequency distribution home range sizes reported for red foxes, coyotes, and wolves.

of mammals, these differences in space use exert a powerful influence on carnivore population structure, affecting their social organization, mating systems, and demography (Bekoff and Daniels 1984; Rubenstein and Wrangham 1986; Clutton-Brock 1989).

1.1. STATISTICAL HOME RANGE ANALYSIS

Estimates of home range size such as those in figure 1.1 come from statistical home range models, which convert spatial distributions of telemetry relocations into an estimate of range size. A number of different models have been developed, including, most famously, the minimum convex polygon method (Odum and Kuenzler 1955), and subsequently, a variety of density estimation

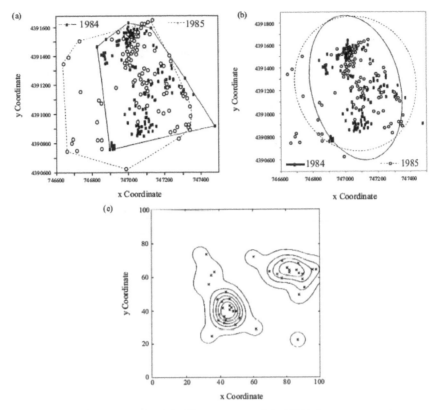

FIGURE 1.2. Examples of statistical home range models. (a) Minimum convex polygon method, (b) bivariate normal method, (c) kernel method. Redrawn from White and Garrott (1997) (panels a and b) and Worton (1989) (panel c).

models such as the bivariate normal (Jennrich and Turner 1969), harmonic mean (Dixon and Chapman 1980), and kernel home range models (Worton 1989)—see MacDonald (1980a), Worton (1987), and Kernohan et al. (2001) for reviews. Examples of these techniques are shown in figure 1.2.

While statistical home range models such as those depicted in this figure provide a useful way to summarize telemetry data, their descriptive nature means that the models have no theoretical or predictive value. Relocation datasets contain a wealth of fine-scale information on the precise spatial and temporal sequence of movements by individuals in relation to their environment and conspecifics. However, virtually all of this detailed biological information is lost when the observations are summarized into aggregate measures of home range size and home range overlap calculated using a statistical home range model (though see Marzluff et al. 2001). As a result, it has been impossible to quantify the underlying determinants of animal space use patterns in anything other than the most coarse-grained manner.

   Moreover, statistical home range models are problematic even as descriptors, producing widely differing portraits of an animal's home range and varying estimates of home range size and overlap (Schoener 1981; Samuel et al. 1985). Issues include the treatment of outliers in minimum convex polygon methods, the long tails of the bivariate normal distribution, the problem of negative values in harmonic mean estimates, and appropriate levels of smoothing in kernel methods (Schoener 1981; Worton 1989; Kernohan et al. 2001). However, since the distributions used to characterize the data are purely statistical, there is no biological basis for preferring one statistical home range model over another.

   In the 1980s an alternative to conventional statistical home range analysis emerged in the form of resource selection analysis (Johnson 1980). In contrast to the descriptive but spatially explicit approach of traditional home range models, resource selection analysis takes a phenomenological, spatially implicit approach to analyzing patterns of space use, identifying habitats and areas used disproportionately in relation to their availability. These analyses can be conducted at a variety of spatial scales including the scale of an individual's home range, sometimes referred to as "third-order" selection (Erickson et al. 2001). As results from a number of studies have shown, the phenomenological, "frequentist" approach of resource selection analysis provides a framework for identifying associations between relative space use by individuals and different forms of environmental spatial heterogeneity such as habitat type, topography, and resource availability (Manly et al. 1993; Cooper and Millspaugh 2001; Erickson et al. 2001). However, the phenomenological and spatially implicit nature of resource selection models limits their predictive capability and their ability to make full use of the biologically rich information contained within telemetry datasets.

1.2. MECHANISTIC HOME RANGE ANALYSIS

A promising alternative to both conventional statistical home range analysis and resource selection analysis is the development of mechanistic home range models. The origins of these models lie in the mathematical analysis of correlated random walks, in which the motion of individual animals is characterized as a sequence of movements at different speeds, orientations, and turning frequencies (Skellam 1951; Skellam 1973; Okubo 1980; Okubo and Levin 2001). The term "correlated random walk" indicates that the locations of an individual are correlated in time and that the individual's rules of movement are stochastic, specified in terms of probability distributions of movement directions, speeds, and rates of turning. This approach has been used to study the movements of cells and microorganisms (Berg and Brown 1974; Alt 1980; Berg 1993; Othmer and Stevens 1997; Hill and Hader 1997; Anderson and Chaplain 1998;

Palsson and Othmer 2000) and insects (Kareiva and Shigesada 1983; Marsh and Jones 1988; Turchin 1998); however, using correlated random walks to study the movements of vertebrates is still in its infancy (Couzin and Krause 2003, though see Gueron and Levin 1993).

The models are mechanistic in the sense that the pattern of space use by an animal is calculated by an explicit mathematical scaling of these underlying rules of movement. Thus unlike instatistical home range models, the patterns of space use obtained from mechanistic home range models are not arbitrary distributions but rather reflect the pattern of space use that results from the underlying set of rules governing the individual's movement, which may incorporate responses to both local and non-local orientation cues (Okubo 1980; Levin and Pacala 1996). This reductionist methodology captures the biological reality that the spatial distribution of relocations in each telemetry dataset shown in figure 1.2 is a macroscopic pattern, the net result of a vast number of movement decisions in response to a variety of environmental and social factors that have influenced the animal as it traversed the landscape.

As we illustrate in this book, mechanistic home range models offer a way to directly integrate theoretical and empirical investigations of animal home range studies. Using this framework, it is possible to formulate models that reflect different hypotheses for the ecological and social factors influencing movement behavior, examine how these alter patterns of space use, then test their predictions against both fine-scale, high-frequency relocation data and, at larger scales, against long-term relocation datasets. In recognizing the macroscopic, scaled nature of home range patterns, mechanistic home range models allow for a more comprehensive use of empirical home range data than statistical home range models. More fundamentally, they also provide the necessary theoretical framework for developing a predictive theory of animal space use.

Through the course of this book we develop and analyze a series of mechanistic home range models for carnivores. We begin in chapter 2 by describing the mathematical procedures for formulating mechanistic home range models from stochastic, individual-based models of animal movement behavior. We then use these methods in chapter 3 to derive a simple mechanistic home range model in which individuals preferentially move in the direction of a home range center, and analyze the properties and predictions of this model, comparing the model's predictions to datasets of red fox and coyote relocations. In chapter 4, we develop an alternative model formulation that incorporates a conspecific avoidance response to foreign scent marks, and compare the fit of this model to the coyote relocation dataset with the fit obtained with the simple model developed in the previous chapter. Chapter 5 examines the properties of the conspecific avoidance model in more detail, using numerical simulations to explore how patterns of space use and scent marks are affected by the movement and behavior of individuals, population density, and location of neighboring home ranges. The qualitative properties of the conspecific avoidance model are then analyzed in chapter 6 by applying analytical methods to a simplified one-dimensional version of the model. In chapter 7, the conspecific avoidance model is extended to incorporate the effects of spatial heterogeneity, taking into consideration such features as terrain characteristics and resource density, and test these more detailed model formulations against observations of coyotes in Yellowstone National Park. The next three chapters are more speculative. In chapter 8, we explore a mechanistic home range model formulation for carnivores in which home ranges arise in the absence of a den site or core area. We then turn our attention to secondary ecological interactions in chapter 9, considering how home range patterns in carnivore populations can have community-level consequences, influencing the spatial distribution of both prey and competitors. In chapter 10, we revisit the localizing tendency model introduced in chapter 3 and establish some connections between the model's predictions and classical statistical measures of space use such as the minimum convex polygon and mean squared displacement. In chapter 11, we illustrate how mechanistic home range models can be combined with game theory to examine the functional significance of different movement strategies and determine evolutionary stable patterns of space use. Finally, in chapter 12, we outline some avenues for future research and discuss some of the broader implications of the mechanistic approach to home range analysis. Mathematical details of calculations are given in appendices A through H.

CHAPTER TWO

From Individual Behavior to Patterns of Space Use

This chapter illustrates the mathematical techniques used to derive the mechanistic home range models analyzed in the later chapters. Our goal is not to provide a comprehensive treatment of how to develop macroscopic equations for animal movement; for this we refer the reader to two excellent general texts on this topic by Okubo (1980) and Turchin (1998). Rather, our goal is to show how the home range models developed in this book are formulated from underlying descriptions of individual movement and interaction behavior.

The model formulation procedure can be broken down into two steps. First, specify the underlying model describing the fine-scale movement behavior of individuals. Second, translate this individual-centered description of movement behavior into a place-centered description of the resulting patterns of space use. Step 1 involves identifying the aspects of the animal's behavior important in governing its space use and encoding these into a set of probabilistic movement and interaction rules for individuals. Since the behaviors most relevant for space use will vary across species and environments, this step requires knowledge of the animal's natural history. Step 2, the subject of this chapter, is more mathematical in nature, and involves formulating equations that translate this individual-centered (Lagrangian) description into a place-centered (Eulerian) description of patterns of space use. As we shall see, the mathematical connection between random movement and diffusion and between directed movement and advection means that patterns of space use for individuals can frequently be approximated using advection-diffusion equations. Our presentation of this material is designed to be as accessible as possible, but the derivation of these equations is inevitably mathematical in nature. Non-mathematical readers are invited to skip to section 2.4, where we summarize the principal results of the analysis.

2.1. MOVEMENT IN ONE DIMENSION

We start by considering an individual moving in a single space dimension x whose movements are uncorrelated in time and space. Biologically, this means that the animal's movement behavior is governed solely by its responses to its current environment and is not contingent on its previous movements. We will revisit this issue again later in the chapter. To describe the expected location x of the animal at each time t, we use a probability density function $u(x, t)$, where the probability that the individual is between points a and b is given by the area under the curve $u(x, t)$ between $x = a$ and $x = b$ (figure 2.1). In other words, if the location of the individual at time t is denoted by a random variable $X(t)$, then

$$\text{Prob}\{a \leq X(t) < b\} = \int_a^b u(x, t)\, dx. \tag{2.1}$$

With the probability density function defined, our task is now reduced to predicting how $u(x, t)$ evolves over time. One way to do this is by assuming that an individual makes a series of discrete movement steps, then deducing how these movements change $u(x, t)$. We start by considering an individual moving on a line divided into small intervals of size dx. When dx is very small, $u(x, t)dx$ is the probability that an individual is located between x and $x + dx$ at time t.

We initially assume that at each time step the individual can only move to the immediately adjacent intervals. We then formulate a so-called *master equation* describing all the possible ways an individual can arrive at interval $[x, x + dx)$[1]

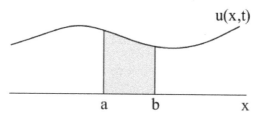

FIGURE 2.1. The probability that the individual is between points a and b is defined to be the area under the curve $u(x, t)$ between $x = a$ and $x = b$.

---

[1]Technically, in the expression $[x, x + dx)$, the left square bracket [indicates that the point x is included, while the right round bracket) indicates that the point $x + dx$ is not included. This allows us to assign each point in space unequivocally to a discrete interval, the next interval begin $[x + dx; x + 2dx)$, and so on.

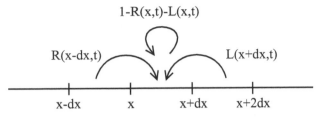

1-R(x,t)-L(x,t)

R(x-dx,t)   L(x+dx,t)

x-dx   x   x+dx   x+2dx

FIGURE 2.2. Over the previous time interval of length τ, an individual on the interval $[x, x + dx)$ at time $t + \tau$ can have arrived from the left, can have remained in the interval, or can have arrived from the right. These terms are given in equation (2.2).

in a single time step:

$$u(x, t + \tau) = R(x - dx, t)u(x - dx, t)$$
$$+ (1 - R(x, t) - L(x, t))u(x, t)$$
$$+ L(x + dx, t)u(x + dx, t). \tag{2.2}$$

Here $R(x, t)$, $L(x, t)$, and $1 - R(x, t) - L(x, t)$ indicate the probabilities of moving to the right, moving to the left, and not moving, in a single time step of length τ (figure 2.2).

To translate this master equation into an equation describing the change in $u(x, t)$ with time, we expand the terms in equation (2.2) using a Taylor series:

$$R(x - dx, t)u(x - dx, t) = R(x, t)u(x, t) - dx \frac{\partial}{\partial x}(R(x, t)u(x, t))$$
$$+ \frac{dx^2}{2}\frac{\partial^2}{\partial x^2}(R(x, t)u(x, t)) + \text{h.o.t.} \tag{2.3}$$

$$L(x + dx, t)u(x + dx, t) = L(x, t)u(x, t) + dx \frac{\partial}{\partial x}(L(x, t)u(x, t))$$
$$+ \frac{dx^2}{2}\frac{\partial^2}{\partial x^2}(L(x, t)u(x, t)) + \text{h.o.t.} \tag{2.4}$$

where h.o.t. represents higher order terms, terms of size $(dx)^3$ and smaller. Because we consider the limiting case where the step length dx becomes small, these higher order terms can be dropped. Substitution into equation (2.2) yields

$$u(x, t + \tau) - u(x, t) \approx -dx \frac{\partial}{\partial x}(u(x, t)(R - L))$$
$$+ dx^2 \frac{\partial^2}{\partial x^2}(u(x, t)(R + L)/2). \tag{2.5}$$

The accuracy of the above equation increases as the space intervals and time steps become small. Dividing equation (2.5) through by τ, taking the limit $dx, \tau \rightarrow 0$, and using the definition of the time derivative $\partial u / \partial t = \lim_{\tau \rightarrow 0}[u(x, t + \tau) - u(x, t)]/\tau$, yields the following partial differential equation:

$$\frac{\partial u}{\partial t} = -\frac{\partial}{\partial x}(c(x, t)u) + \frac{\partial^2}{\partial x^2}(d(x, t)u). \tag{2.6}$$

The coefficients c and d are given by

$$c = \lim_{dx, \tau \rightarrow 0} dx \frac{(R - L)}{\tau} \quad \text{and} \quad d = \lim_{dx, \tau \rightarrow 0} dx^2 \frac{(R + L)}{2\tau} \tag{2.7}$$

respectively.

When combined with appropriate initial and boundary conditions, describing the initial location of the individual and what happens at the edges of the region, equation (2.6) predicts the time-dependent pattern of space use of the individual, defined in terms of a time-varying probability density function $u(x, t)$ for the expected location of the individual. The equation has an advection parameter c with units of distance over time, and a diffusion parameter d with units of distance squared over time.[2]

Variable Step Length

A generalization of the above movement model allows individuals to jump more than one space step in each time step. To define this more general process, we suppose that the individual is found on one of the small intervals $[x', x' + dx)$ at time t, and we assign a probability to the event that the individual is found at each interval $[x, x + dx)$ at time $t + \tau$. This is done through a *redistribution kernel* $k(x', x, \tau, t)$. Here $k(x', x, \tau, t)dx'dx$ is the probability of moving from $[x', x' + dx')$ at time t to $[x, x + dx)$ at time $t + \tau$. The redistribution kernel k describes the potential movement of the individual in both the positive $(x > x')$ and negative $(x < x')$ directions over the time interval (figure 2.3).

The conservation equation (2.2) describing movement of the individual over a single time step becomes

$$u(x, t + \tau) = \sum_{n=-\infty}^{\infty} u(ndx', t)k(ndx', x, \tau, t) \, dx'. \tag{2.8}$$

---

[2]Note that in a model with a bias in the random walk (i.e., $R \neq L$) we must define the *bias per unit length* b so that it converges to a fixed bias speed as the space step dx and the time step τ approach zero. A simple case incorporating the bias per unit length uses the probability of moving to the right as $R = 1/2 + bdx$ and of moving to the left as $L = 1/2 - bdx$. The above formulas yield $d = \lim_{dx, \tau \rightarrow 0} dx^2/(2\tau)$ and $c = 2bd$.

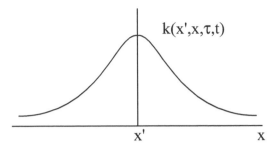

FIGURE 2.3. The redistribution kernel $k(x', x, \tau, t)$ describes the probability that an individual located at x' $x' + dx$ at time t will have moved to a position between $[x, x + dx)$ x $x + dx$ at time $t + \tau$.

This generalization of equation (2.2) covers the possibility of jumps of any size in a single time step. The limiting case where the space step becomes small $(dx \to 0)$ changes the right-hand side of this equation, which becomes an integral, yielding

$$u(x, t + \tau) = \int_{-\infty}^{\infty} u(x', t) k(x', x, \tau, t) dx'. \tag{2.9}$$

The above equation is translated into a differential equation using the same procedure used for equation (2.2). Expanding the terms on the right-hand side using a Taylor series in x' and t and then considering the limit $dx, \tau \to 0$, we get

$$\frac{\partial u(x, t)}{\partial t} = -\frac{\partial}{\partial x}[c(x, t)u(x, t)] + \frac{\partial^2}{\partial x^2}[d(x, t)u(x, t)]. \tag{2.10}$$

Details of the derivation can be found in appendix A. This equation is the same as equation (2.6) but with advection and diffusion coefficients now given by

$$c(x, t) = \lim_{\tau \to 0} \frac{1}{\tau} \int_{-\infty}^{\infty} (x' - x) k(x', x, \tau, t) dx' \quad \text{and}$$

$$d(x, t) = \lim_{\tau \to 0} \frac{1}{2\tau} \int_{-\infty}^{\infty} (x' - x)^2 k(x', x, \tau, t) dx'. \tag{2.11}$$

These are the first and second moments of the distribution of movement distances predicted by the redistribution kernel as $\tau \to 0$, referred to as the first and second *infinitesimal moments* of k. An alternative approach to deriving equation (2.6) is discussed in appendix B.

2.2. MOVEMENT IN TWO DIMENSIONS

The above derivation readily generalizes to the more biologically relevant case
of movement in two dimensions.[3] We define $u(\mathbf{x}, t)$ to be the two-dimensional
probability density function for the location of an individual at time t where $\mathbf{x}$
is a vector indicating the (x, y) position of the individual. In two dimensions,
equation (2.9) becomes

$$u(\mathbf{x}, t + \tau) = \int_{-\infty}^{\infty} \int_{-\infty}^{\infty} u(\mathbf{x}', t) k(\mathbf{x}', \mathbf{x}, \tau, t) \, dx \, dy \qquad (2.12)$$

where $k(\mathbf{x}', \mathbf{x}, \tau, t) d\mathbf{x}' d\mathbf{x}$ is the probability of moving from a small rectangle $d\mathbf{x}'$
located at $\mathbf{x}'$ at time t to a small rectangle $d\mathbf{x}$ located at $\mathbf{x}$ at time $t + \tau$. As in
equation (2.9), using Taylor expansion of equation (2.12) and then taking the
limit as τ becomes small, yields

$$\frac{\partial u}{\partial t} + \nabla \cdot [\mathbf{c}(\mathbf{x}, t) u] = \frac{\partial^2 \left(d_{xx}(\mathbf{x}, t) u \right)}{\partial x^2} + \frac{\partial^2 \left(d_{xy}(\mathbf{x}, t) u \right)}{\partial x \partial y}$$

$$+ \frac{\partial^2 \left(d_{yx}(\mathbf{x}, t) u \right)}{\partial y \partial x} + \frac{\partial^2 \left(d_{yy}(\mathbf{x}, t) u \right)}{\partial y^2} \qquad (2.13)$$

where the advection term

$$\mathbf{c}(\mathbf{x}, t) = \lim_{\tau \to 0} \frac{1}{\tau} \int (\mathbf{x}' - \mathbf{x}) k(\mathbf{x}', \mathbf{x}, \tau, t) \, d\mathbf{x}' \qquad (2.14)$$

is a vector. ∇ denotes the gradient or spatial derivative operator in two spatial
dimensions $\nabla = (\partial/\partial x, \partial/\partial y)^T$ (the T here means transpose), and

$$d_{xx}(\mathbf{x}, t) = \lim_{\tau \to 0} \frac{1}{2\tau} \int (x' - x)^2 k(\mathbf{x}', \mathbf{x}, \tau, t) d\mathbf{x}'$$

$$d_{xy}(\mathbf{x}, t) = \lim_{\tau \to 0} \frac{1}{2\tau} \int (x' - x)(y' - y) k(\mathbf{x}', \mathbf{x}, \tau, t) d\mathbf{x}'$$

$$d_{yx}(\mathbf{x}, t) = d_{xy}(\mathbf{x}, t)$$

$$d_{yy}(\mathbf{x}, t) = \lim_{\tau \to 0} \frac{1}{2\tau} \int (y' - y)^2 k(\mathbf{x}', \mathbf{x}, \tau, t) d\mathbf{x}' \qquad (2.15)$$

are generalized diffusion terms. The first and last equations of (2.15) are the
infinitesimal second moments of the redistribution kernel in the x and y directions as τ becomes arbitrarily small. These terms are always non-negative.
The middle two equations describe possible covariances (scaled correlations)

---

[3]Higher dimensions also follow in a similar manner.

between movements in the x and y directions as τ becomes arbitrarily small, and can be either positive or negative.

A standard simplification of equation (2.15) assumes that the covariances between movements in the x and y directions are negligible and that the infinitesimal second moments are the same in each direction. This *isotropic diffusion* assumption means that the second and third equations of (2.15) are both zero and the first and last equations are equal, which simplifies (2.10) to the following equation for the time-dependent pattern of space use by the individual:

$$\frac{\partial u}{\partial t}(\mathbf{x}, t) = -\nabla \cdot (\mathbf{c}(\mathbf{x}, t)u) + \nabla^2 (d(\mathbf{x}, t)u) \tag{2.16}$$

where $\nabla^2 u = \nabla \cdot \nabla u = \partial^2 u / \partial x^2 + \partial^2 u / \partial^2 y$. Further details can be found in Bharucha-Reid (1960).

2.3. DIRECTED AND RANDOM MOTION

The relationship between the parameters d and c in equations (2.10) and (2.16) and the animal's underlying movement behavior can be highlighted by considering two extreme forms of movement: directed motion and random motion.

Directed Motion

Suppose an individual moves in a fixed direction at a constant speed. If its spatial position is recorded at fixed time intervals, this results in a series of equally spaced relocations that lie in a straight line (figure 2.4a). The direction and distance between the individual's current and subsequent recorded position is always given by $\gamma\tau\vec{\mathbf{x}}$, where γ is the speed of movement, τ is the time interval between relocations, and $\vec{\mathbf{x}}$ is a unit vector pointing in the direction of movement. In mathematical terms, this corresponds to a redistribution kernel k that, when plotted in the space $\mathbf{x} - \mathbf{x}'$, is a delta function[4] situated at $\gamma\tau\vec{\mathbf{x}}$,

$$k(\mathbf{x}', \mathbf{x}, \tau, t) = \delta(\mathbf{x} - \mathbf{x}' - \gamma\tau\vec{\mathbf{x}}) \tag{2.17}$$

(Figure 2.4b).

---

[4]A delta function $\delta(\mathbf{x} - \mathbf{x}_0)$ is a distribution which integrates to one, but has its entire mass concentrated at the point $\mathbf{x}_0$. This gives it the rather useful property that $\int f(\mathbf{x})\delta(\mathbf{x}-\mathbf{x}_0)\, d\mathbf{x} = f(\mathbf{x}_0)$ for any smooth function f.

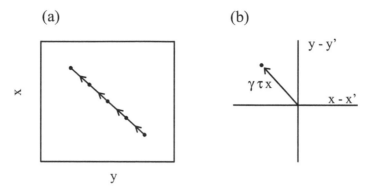

FIGURE 2.4. (a) Trajectory and relocations of an individual exhibiting pure directed motion at a constant speed γ in a single direction $\vec{\mathbf{x}}$. Line shows the path of the individual. Points show the recorded locations $(x_1, y_1), (x_2, y_2), \ldots (x_n, y_n)$ of the individual recorded at successive time intervals $i - 1, i, i + 1$. (b) When plotted in the space $(x - x', y - y')$, the redistribution kernel k resulting from the directed movement shown in (a) is a delta function located at $\gamma\tau\vec{\mathbf{x}}$ (see eq. (2.17)).

Substituting equation (2.17) into equations (2.14) and (2.15), integrating over x and y, and using the properties of the delta function (see footnote) yields

$$\mathbf{c}(\mathbf{x}, t) = \lim_{\tau \to 0} \frac{1}{\tau} \int (\mathbf{x'} - \mathbf{x})\delta(\mathbf{x'} - \mathbf{x} - \gamma\tau\vec{\mathbf{x}}) \, d\mathbf{x'}$$

$$= \lim_{\tau \to 0} \frac{1}{\tau} \left[\gamma\tau\vec{\mathbf{x}} \right] \qquad (2.18)$$

and

$$d_{xx}(\mathbf{x}, t) + d_{yy}(\mathbf{x}, t) = \lim_{\tau \to 0} \frac{1}{\tau} \int |\mathbf{x'} - \mathbf{x}|^2 \delta(\mathbf{x'} - \mathbf{x} - \gamma\tau\vec{\mathbf{x}}) \, d\mathbf{x'}$$

$$= \lim_{\tau \to 0} \frac{1}{2\tau} \left[\gamma^2\tau^2 \right] \qquad (2.19)$$

which in the limit $\tau \to 0$ gives

$$\mathbf{c}(\mathbf{x}, t) = \gamma\vec{\mathbf{x}} \quad \text{and}$$

$$d_{xx}(\mathbf{x}, t) = d_{yy}(\mathbf{x}, t) = 0. \qquad (2.20)$$

A similar argument can be used to show that $d_{xy}(\mathbf{x}, t) = 0$.

Thus, directed movement by the individual results in a non-zero advection term in equation (2.16). By substituting equation (2.20) into equation (2.16), we get

$$\frac{\partial u}{\partial t} = \underbrace{-\nabla \cdot (\mathbf{c}(\mathbf{x}, t)u)}_{\text{directed movement}} = -\nabla \cdot \left(\gamma \vec{\mathbf{x}} u\right) = \gamma \vec{\mathbf{x}} \cdot \nabla u,$$

(2.21)

a pure advection equation. Given an initial probability density function for the location of the animal $u(x, 0) = u_0(x)$, the solution to equation (2.21) is

$$u(\mathbf{x}, t) = u_0(\mathbf{x} - \gamma \vec{\mathbf{x}} t).$$

(2.22)

Readers are invited to verify this by differentiating equation (2.22) with respect to t. Thus, we see that the expected location of the individual at any time t is simply the probability density function for the initial location of the individual u_0, frame-shifted a distance γt in direction $\vec{\mathbf{x}}$. We may have guessed this without ever formulating an advection-diffusion equation. However, the advection-diffusion formulation becomes invaluable when, as in the next section, individuals also have a random component in their fine-scale movement behavior.

Random Motion

Now consider an individual moving at a constant speed that changes its direction at random intervals with no preferred direction of movement (figure 2.5). Following the same methodology as for the directed movement case, the individual's spatial position is recorded at fixed time intervals τ, resulting in a series of relocations. In the case of directed movement, the relocations captured the actual movement path of the individual regardless of their frequency (see figure 2.4a). However, in the case of random motion, the relocations only approximate the actual path of the individual. When the interval between successive relocations is small, the pattern of relocations closely resembles its true path (figure 2.6a), but when the time between successive relocations is longer, the large-scale pattern of movement is still captured, but much of the fine-scale detail is lost (figure 2.6b).

While high-frequency relocations such as those in figure 2.6a more closely follow the true path of the individual, the capturing of this fine-scale detail comes at a cost. When the time between relocations is short, the movement directions implied by successive relocations become correlated. This arises because even though the individual moves in random directions, in many instances the individual has not changed direction during the time between successive relocations. These correlations are problematic because they violate an assumption we made

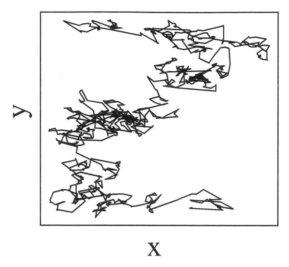

FIGURE 2.5. Trajectory of an individual exhibiting random motion in two space dimensions (x, y). The individual moves at constant speed but changes direction at random time intervals, with no preferred direction of movement.

when deriving equations (2.10) and (2.16), namely, that successive movement directions of the individual are uncorrelated. Such serial autocorrelation can be explicitly incorporated into macroscopic equations for space use, though at the cost of increased model complexity. This issue is discussed in more detail in appendix C.

Provided that the autocorrelation is not itself of interest (as is often the case), an alternative, simple approach avoids violating the independence assumption by approximating the individual's trajectory with a set of relocations sufficiently separated in time so that the correlation in movement direction between successive observations is near zero (Turchin 1998). For example, while subsequent movement directions implied by the relocations in figure 2.6a are highly correlated (see figure 2.7a), the lower frequency relocations shown in figure 2.6b imply subsequent movement directions that are relatively uncorrelated (see figure 2.7b).

It is important to note that the assumption that *movement directions* are uncorrelated between successive observations is quite distinct from the assumption made when fitting statistical models to relocation data—that the *spatial locations* are uncorrelated between successive observations. The difference has practical significance in that fairly short time intervals τ ensure the former, whereas long time intervals between relocations are needed to ensure the latter.

When using this approximation procedure, rather than taking the limit $\tau \to 0$ when calculating d and c in equations (2.14) and (2.15), the coefficients are estimated using the uncorrelated relocations. A heuristic explanation of this

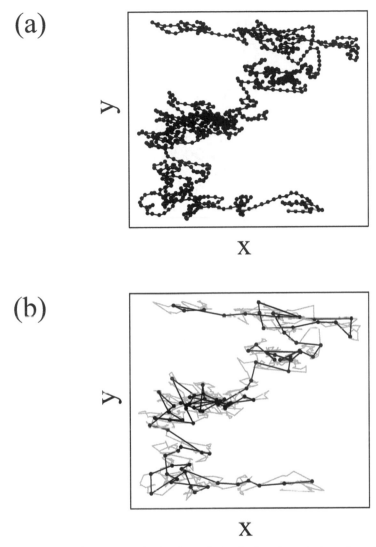

FIGURE 2.6. (a) The pattern of relocations obtained when the trajectory shown in figure 2.5 is sampled at two different frequencies. If the time between successive relocations is short as in (a), the relocations (–•–) capture the detailed trajectory of the individual seen in figure 2.5. When the time between successive relocations is longer as in (b), the relocations (–•–) capture only the large-scale components of the individual's actual trajectory (shown in the panel as gray lines).

approach is that in doing so, we are approximating the individual's actual movement behavior with a stochastic movement process that captures its broad-scale pattern of movement but, unlike its actual movements, is uncorrelated at fine scales.

(a)

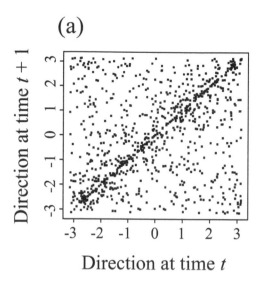

Direction at time t

(b)

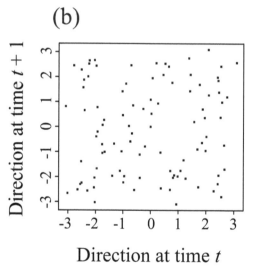

Direction at time t

FIGURE 2.7. Correlation in successive movement directions implied by the relocations shown in figure 2.6. (a) When the trajectory is sampled at high frequency, as in figure 2.6a, the individual's current movement direction exhibits a high degree of correlation with its prior movement direction. (b) When the trajectory is sampled at low frequency as in figure 2.6b, the individual's current movement direction is uncorrelated with its previous movement direction.

As in the previous case, we calculate the reorientation kernel k by plotting the displacements in the x and y directions that occur with movement from location (x', y') to (x, y). Since the individual changes its direction at random intervals and moves in random directions, this yields a scatter of movement distances

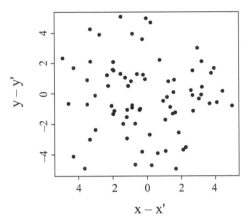

FIGURE 2.8. Distribution of movements obtained from the uncorrelated relocations shown in figure 2.6b provide an estimate for the reorientation kernel k. Plot shows the displacement in the x and y directions, $x - x'$ and $y - y'$, implied by successive relocations $\mathbf{x}_i - \mathbf{x}_{i-1}$. The average displacement is zero; however, the average squared displacements $(x - x')^2$ and $(y - y')^2$ remain positive.

and directions (figure 2.8). Considering first the advection coefficient, $\mathbf{c}(\mathbf{x}, t)$, our estimate for this term is the average distance moved between successive relocations per unit time $\mathbf{x}_{i+1} - \mathbf{x}_i$ (see figure 2.8):

$$\mathbf{c}(\mathbf{x}, t) = \frac{1}{\tau} \frac{1}{(n_r - 1)} \sum_{i=1}^{n_r - 1} (\mathbf{x}_{i+1} - \mathbf{x}_i). \qquad (2.23)$$

Since the individual has no preferred direction of movement, provided a sufficient number of relocations have been taken, the average distance between successive relocations $\mathbf{x}_{i+1} - \mathbf{x}_i$ will be zero, i.e.,

$$\mathbf{c}(\mathbf{x}, t) = 0. \qquad (2.24)$$

The estimate for the diffusion term $d(\mathbf{x}, t)$ is the average squared distance moved between successive locations in the x and y directions per unit time, i.e.,

$$d_{xx}(\mathbf{x}, t) = \frac{1}{2\tau} \frac{1}{(n_r - 1)} \sum_{i=1}^{n_r - 1} (x_{i+1} - x_i)^2,$$

$$d_{yy}(\mathbf{x}, t) = \frac{1}{2\tau} \frac{1}{(n_r - 1)} \sum_{i=1}^{n_r - 1} (y_{i+1} - y_i)^2. \qquad (2.25)$$

(a)

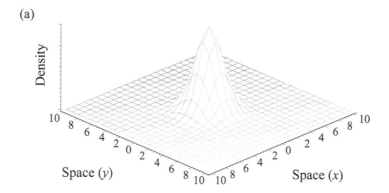

(b)

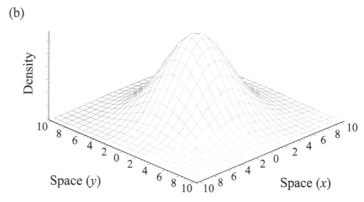

FIGURE 2.9. When a single individual is released from the origin at time $t = 0$, the probability density function for its initial expected location is a delta function at origin (see section 2.4). Equation (2.26) then describes how this probability density function changes with time. Its solution (equation 2.27) is a spreading Gaussian distribution, which describes the growing lack of certainty of the location of the individual due to its random movement. In the example shown, the diffusion coefficient d is 1, and the solution is shown for $t = 0.5$ (a) and $t = 3.0$ (b).

In contrast to the average displacement distance, the average squared displacement distance remains positive, i.e., $d > 0$.

Thus, in the case of random movement, the macroscopic equation for space use, equation (2.16), reduces to

$$\frac{\partial u}{\partial t} = \underbrace{\nabla^2 (d(\mathbf{x},t)u)}_{\text{random movement}} \quad , \tag{2.26}$$

a pure diffusion equation.

To illustrate the qualitative behavior of solutions to equation (2.26), we consider the case where d is a constant and the individual is released from

$\mathbf{x} = (0,0)$ at time $t = 0$. The solution to equation (2.26) is a bivariate Gaussian distribution, with mean zero and variance $4dt$ (Murray 1989):

$$u(\mathbf{x}, t) = \frac{\exp(-(x^2 + y^2)/(8dt))}{8dt\pi}. \tag{2.27}$$

The mathematical details of how to obtain this solution can be found in Haberman (1987), but the reader may verify the solution by taking partial derivatives of equation (2.27) twice with respect to each space variable, and once with respect to time, to obtain equation (2.26). The spreading Gaussian distribution given by eq. (2.26) describes the growing lack of certainty of the location of the individual due to its random movements (figure 2.9). The mixture of directed movement (advection) and random movement (diffusion) is discussed in the following chapter in the context of home range pattern formation.

2.4. PREDICTING HOME RANGE PATTERNS

The above cases imply the following interpretation of the space use equation (2.16):

$$\frac{\partial u}{\partial t} = \underbrace{-\nabla \cdot (\mathbf{c}(\mathbf{x}, t)u)}_{\substack{\text{directed motion} \\ \text{(advection)}}} + \underbrace{\nabla^2 (d(\mathbf{x}, t)u)}_{\substack{\text{random motion} \\ \text{(diffusion)}}} . \tag{2.28}$$

Equation (2.28) predicts the changing pattern of space use $u(\mathbf{x}, t)$ by an individual. Throughout most of this book we will be interested in determining the equilibrium solution of the space use equation, which we calculate by setting the time derivative of the above equation equal to zero. This concept of a home range as a steady-state pattern of space use expressed in terms of a probability density function closely matches the modern concept of an animal's home range (Worton 1987).

The pattern of space use predicted by equation (2.28) is affected by the form of the boundary conditions for equation (2.28) that describe the behavior of the individual at the edges of the spatial region under consideration. In this book we will primarily be using so-called "zero-flux" boundary conditions that reflect animals moving in a finite, self-contained region. In a few instances, we use "zero at infinity" boundary conditions that reflect individuals moving in a region that has no edges. For a general discussion of boundary conditions with regard to animal movement, see Turchin (1998).

When considering time-dependent patterns of space use, what is known about the location of the individual at the beginning of the time period is specified as an initial condition. This takes the form of a probability density function $u_0(x)$ that describes the expected location of the individual at $t = 0$ (i.e., $u(x, 0) = u_0(x)$). For example, if the individual is known to be at a specific point, then $u_0(x)$ would be a delta function at this location. Conversely, if the initial location of the individual is not known, then another spatial distribution such as a uniform distribution might be an appropriate initial condition for the simulation.

2.5. SUMMARY

In this chapter we illustrate the mathematical techniques used to derive equations for the expected pattern of space use that results from an underlying model of an individual's movement behavior. The starting point for the derivation is a description of the individual's movement behavior, specified as a stochastic movement process consisting of a series of movements of different lengths, durations, and directions. We show how the outcome of such stochastic movement processes can be approximated by advection-diffusion equations (ADEs). We then analyze two special cases that highlight the relationship between different types of movement and the resulting equations for space use. Advection describes directed components of an individual's movement, while diffusion describes the random, stochastic aspects of its movement behavior (see eq. (2.28)). The form of the equation's coefficients and how they vary as a function of the animal's spatial position is governed by the underlying rules of movement. We illustrate this in more detail in the next chapter by formulating a simple mechanistic home range model.

CHAPTER THREE

A Simple Mechanistic Home
Range Model

Figure 3.1 shows the spatial extent of relocations of two carnivores, a wolf and a coyote, as a function of time from their first relocation. Initially, their space use increases rapidly, but as the sampling continues, the spatial extent of the relocations saturates, indicating that both individuals are restricting their movements to particular areas. This saturation is not consistent with random movement by these individuals, an assertion we will later prove in chapter 10, and indicates the presence of a localizing tendency in their movement behavior. This is common in many carnivore species where the need to provision offspring means that den sites routinely act as focal points for the movements of adults, especially during periods of breeding and pup rearing (Bekoff and Wells 1982; Moehlman 1986). The above observations form the basis for what is arguably the simplest mechanistic home range model, proposed first by Holgate

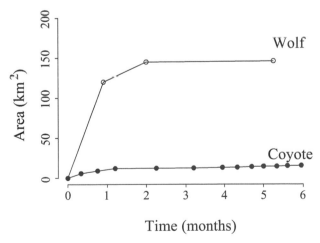

FIGURE 3.1. Increase in the areal extent of wolf ($-\circ-$) and coyote ($-\bullet-$) relocations as a function of the length of radio tracking. Data re-plotted from Messier and Barrette (1982) and Okarma et al. (1998).

(1971) then subsequently by Okubo (1980), and henceforth referred to as the "localizing tendency" model.[1]

3.1. MODEL OF INDIVIDUAL MOVEMENT BEHAVIOR

As we saw in the previous chapter, the starting point for developing a mechanistic home range model is to specify a set of movement rules for individuals in the form of a redistribution kernel describing the probability of an individual moving from one location to another as a function of time and its current spatial position. At the scale of individual movement decisions, we can express the localizing tendency in the form of a non-uniform distribution of movement directions

$$\underbrace{K(\phi, \widehat{\phi})}_{\text{distribution of movement directions}} \qquad , \qquad (3.1)$$

where $K(\phi, \widehat{\phi})d\phi$, the probability of moving in direction ϕ, is biased toward a particular location or area whose direction is indicated by the angle $\widehat{\phi}$.

A convenient functional form for K is a *von Mises* distribution, a circular distribution that plays a similar role to the normal distribution in linear statistics (Batschelet 1981).[2] It is unimodal, with probability density function

$$K(\phi, \widehat{\phi}) = \frac{1}{2\pi I_0(\kappa)} \exp\left[\kappa \cos(\phi - \widehat{\phi})\right], \qquad (3.2)$$

and two parameters, $\widehat{\phi}$ ($-\pi \leq \widehat{\phi} \leq \pi$) and κ ($\kappa \geq 0$). The angle $\widehat{\phi}$ is the mode of the distribution and also the mean direction (figure 3.2). $I_0(\kappa)$ is a modified Bessel function that normalizes K to integrate to 1. The parameter κ is the concentration parameter, which governs the degree of non-uniformity in the distribution of movement directions. When $\kappa = 0$, the distribution is a uniform circular distribution indicating no preferred direction of movement (figure 3.2a). As κ increases above zero, the distribution becomes increasingly concentrated around the mean direction $\widehat{\phi}$ (figure 3.2b–d). We can use the above distribution to specify a turning kernel for individuals that incorporates a localizing tendency. Recalling our analysis in section 2.3 of the previous chapter, we consider the case where an animal is relocated at regular time intervals of length τ, where τ is short enough that the animal is unlikely to have traveled far relative to its overall home range, but long enough to give negligible correlation between successive relocations.

---

[1] This model has its antecedents in the biased random walk models used in statistical physics (Smoluchowski 1916; Kac 1947).

[2] Also referred to as the circular normal distribution (Gumbel 1954).

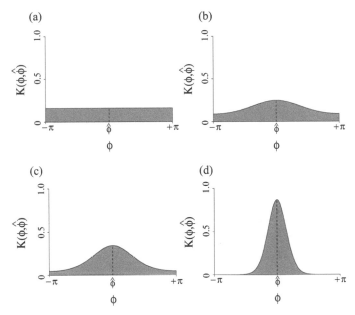

FIGURE 3.2. Distributions of directions resulting from a von Mises distribution (eq. (3.2)) with mean direction $\widehat{\phi} = 0$ and varying levels of the concentration parameter κ. (a) $\kappa = 0$, (b) $\kappa = 0.5$, (c) $\kappa = 1$, (d) $\kappa = 5$.

Suppose that the movements of the individual over time τ can be expressed as a product of two distributions, one describing the distances moved and the other describing the angles of movement. Mathematically, this corresponds to a redistribution kernel of the following form:

$$k(\mathbf{x}, \mathbf{x}', \tau, t) = \underbrace{\frac{1}{\rho} f_\tau(\rho) \cdot K_\tau(\phi, \widehat{\phi})}_{\text{probability density for moving from location } \mathbf{x}' \text{ to } \mathbf{x}} \qquad (3.3)$$

where the kernel k has been translated from Cartesian (x, y) coordinates into polar (ρ, ϕ) coordinates. Here ρ is the distance between the starting point $\mathbf{x}'$ and the finishing point $\mathbf{x}$ ($\rho = |\mathbf{x}' - \mathbf{x}|$), and ϕ is the angle between the starting point and the finishing point, ($\phi = \tan^{-1}(y - y'/x - x')$), where $\widehat{\phi}$ is the direction of the individual's home range center from its current position (figure 3.3).[3]

---

[3] Here the arctan function is extended from its usual range of $(-\pi/2, \pi/2)$ to $(-\pi, \pi]$, by taking into account which quadrant the point $(x - x', y - y')$ is in.

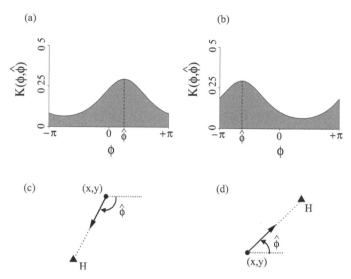

FIGURE 3.3. In the Holgate-Okubo localizing tendency model, an individual's movements are biased in the direction of a home range center or core area. (a) and (b) The
probability distribution of the individual's movement direction $K(\phi, \widehat{\phi})$ (see eq. (3.1)),
varies as a function of the animal's spatial position, at all locations being biased in
the direction of its home range center. In the example shown, ϕ is drawn from a von
Mises distribution (eq. (3.2)), whose mean direction vector points in the direction of
the individual's home range center $\widehat{\phi}$. The degree of non-uniformity in the individual's
movement direction is governed by the concentration parameter κ of the distribution
(see figure 3.2). (c) and (d) The direction of the bias $\widehat{\phi}$ depends on the individual's
position ($\bullet$) relative to its home range center H ($\blacktriangle$) ($\widehat{\phi} = \tan^{-1}(y - y_H/x - x_H)$, where
(x, y) is the current position of the individual and (x_H, y_H) is the location of its home
range center). The direction of movement ϕ is given by $\tan^{-1}(y - y'/x - x')$, where
(x', y') is the prior position of the individual.

$K_\tau(\phi, \widehat{\phi})d\phi$ is the probability of moving in direction ϕ during the time interval of length τ, and the quantity $f_\tau(\rho)d\rho$ is the probability that the individual at
point $\mathbf{x}'$ will move to a point $\mathbf{x}$ that is between distance ρ and $\rho + d\rho$ from $\mathbf{x}'$. The
$1/\rho$ that precedes $f_\tau(\rho)$ in equation (3.3) translates the probability of moving
a given distance and direction into a probability of moving from one area to
another.

We incorporate the localizing tendency of the individual by specifying that
the individual's distribution of movement directions (K_τ) is given by the von
Mises density function (eq. (3.2)) with concentration parameter κ_τ, and $\widehat{\phi}$ is the
angle between the individual and the individual's home range center (x_H, y_H)
as given by $\widehat{\phi} = \tan^{-1}(y - y_H/x - x_H)$ (figure 3.3).

Note that the functions f_τ and K_τ are fixed for a given sampling frequency τ,
but will change if τ is changed.

3.2. CHARACTERIZING THE MOVEMENT BEHAVIOR
OF A RED FOX

We now investigate the ability of the localizing tendency movement model to characterize the movement behavior of a red fox (*Vulpes vulpes*). The dataset we use for this analysis is Siniff and Jessen's (1969) dataset of red fox movements collected using the Minnesota Cedar Creek Long-Term Ecological Research Site (figure 3.4). Relocations were collected between May 5 and June 5, 1969 using an automated tracking system that determined the animal's spatial position every 10 minutes from high accuracy triangulations (0.5° error) recorded by two fixed monitoring stations (see Siniff and Jessen (1969) for details). As can be seen in figure 3.4, the high-frequency relocations capture the detailed move ments of the individual and clearly indicate that it is confining its movements to a particular area (figure 3.4).

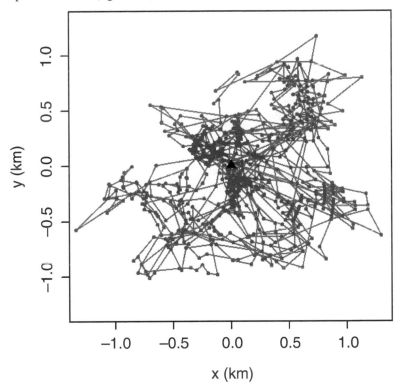

FIGURE 3.4. Gray lines show the observed trajectory of an individual red fox recorded by the Minnesota Cedar Creek tracking system over a 30-day period collected by Siniff and Jessen (1969). The time between successive relocations τ is 10 minutes (0.17 hours). The (x, y) location of the individual is expressed in kilometers from the individual's home range center (▲), assumed to correspond to the centroid of the relocations. The triangulation error is 0.5°.

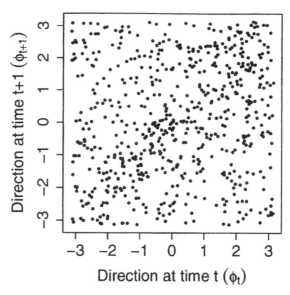

FIGURE 3.5. Relationship between successive movement directions for the trajectory shown in figure 3.4 in which the individual's location is recorded every 10 minutes. The x axis is the individual's movement direction during the current time step (t) and the y axis is the individual's movement direction in the subsequent time interval $(t+1)$. At this sampling frequency, there is no significant correlation in the subsequent movement directions of the individual (corr. coeff. $= 0.18, n.s$).

We can use these high frequency observations to estimate the coefficients of equation (3.3). Recalling our analysis in section 2.3 of chapter 2, we first check the correlation between subsequent movement directions of the individual inferred by the successive relocations of the individual. While relocations in figure 3.4 are sufficiently frequent to have captured much of the fine-scale detail of the individual's movement, the interval of 10 minutes between successive relocations is sufficiently long that there is no significant correlation in subsequent movement directions (figure 3.5). Thus no resampling is needed before using the relocations to estimate the individual's fine-scale–movement behavior, assuming no correlation in subsequent movement directions.

Equation (3.3) consists of two components: the individual's distribution of movement directions relative to its home range center (K_τ) and its distribution of movement distances between successive relocation (f_τ). Figure 3.6a shows the observed distribution of the individual's movement directions around its home range center. Since no data on the individual's den location or core foraging areas was available, we used the centroid of the relocations for the location of the individual's home range center (▲ in figure 3.4). As expected, the distribution of the individual's movement directions shows the presence of a localizing tendency in the fox's movement behavior. The dashed line in the figure shows the fit of equation (3.2) to these observations, which yields an estimate of 0.26

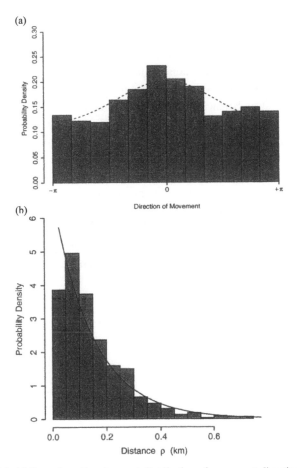

FIGURE 3.6. (a) Bars show the observed distribution of movement directions relative to the location of the home range center $K_\tau(\phi, \widehat{\phi})$ for the relocation dataset shown in figure 3.4. Line shows the fit of a von Mises distribution (eq. (3.2)). The maximum likelihood value for κ_τ, the concentration parameter that describes the magnitude of preferential movement towards the home range center, is 0.26. (b) Observed distribution of distances (ρ) between successive relocations, $f_\tau(\rho)$, for the relocation dataset shown in figure 3.4. The mean distance between successive relocations is $\bar\rho_\tau = 0.14$ km. The parameter b is the bias per unit length traveled $b = \kappa_\tau / \bar\rho_\tau$ (see footnote 2 from previous chapter), yielding an estimate for b of $0.26/0.14 = 1.86$ km^{-1}.

for the concentration parameter κ_τ of the von Mises distribution. For more details on how to calculate κ_τ from the observations, see appendix D.

Figure 3.6b shows the individual's distribution of movement distances between successive relocations $f_\tau(\rho)$. The average distance moved by the individual between relocations $\bar\rho_\tau$ is 0.14 km. The line in this figure is an exponential distribution with the same mean as the observations, indicating that the individual's distribution of movement distances is approximately exponential (figure 3.6b).

3.3. EQUATIONS FOR PATTERNS OF SPACE USE

Heuristically, we might expect that the random movement reflected by the diffusion term in equation (3.5) will cause spatial spread, and the directed movement reflected by the advection term will cause localization around the home range center. The resulting balance between spread and localization will give rise to a restricted pattern of space use or home range for the individual.

An important advantage of formulating the space use equation (3.5) is that we can gain mathematical insight into the relationship between underlying movement behavior and the resulting home range. Recalling our analyses in the previous chapter, we now compute the mean and mean-squared displacement per unit time that result from the above redistribution kernel. In appendix E, we show that, when the distribution of movement distances is exponential with mean $\bar{\rho}_\tau$ and the distribution of movement directions given by a von Mises distribution (equation 3.2) with concentration parameter

$$\kappa_\tau = b\bar{\rho}_\tau, \tag{3.4}$$

inserting equation (3.3) into equations (2.14) and (2.15) results in the following advection-diffusion equation:

$$\frac{\partial u}{\partial t}(\mathbf{x}, t) = \underbrace{d\nabla^2 u}_{\text{random motion}} - \underbrace{c\nabla \cdot (u\vec{\mathbf{x}})}_{\text{directed motion toward } \vec{\mathbf{x}}}, \tag{3.5}$$

where d and c are given by

$$d = \bar{\rho}_\tau^2/4\tau \quad \text{and} \quad c = \kappa_\tau \bar{\rho}_\tau/2\tau. \tag{3.6}$$

This is the same form as equation (2.16) with advection coefficient $\mathbf{c}(\mathbf{x}, t) = c\vec{\mathbf{x}}$, where $\vec{\mathbf{x}}$ is a unit vector pointing toward the home range center and c is the magnitude of the bias in movement direction, and diffusion coefficient $d(\mathbf{x}, t) = d$, where d is a constant reflecting the random aspect of the individual's motion.[4] Inserting into the equation the estimates for κ, $\bar{\rho}$, and τ calculated for the red fox movement data shown in figure 3.5, $\tau = 0.17$ h, $\bar{\rho}_\tau = 0.14$ km, $\bar{\rho}_\tau^2 = 0.28$ km^2 (see figure 3.6), yields estimates of 0.085 km h^{-1} and 0.041 km^2 h^{-1} for c and d respectively.

Now that we have established the connection between the parameters of equation (3.5) and the coefficients of equation (3.3), we can now use

---

[4] Note that, even when K_τ and f_τ are drawn from distributions other than a von Mises distribution and an exponential distribution, a similar procedure to the one described in appendix E will yield an equation like equation (3.5), provided that the distribution of movement directions exhibits a bias toward the den site.

equation (3.5) to predict the pattern of space use that results from the individual's fine-scale movement behavior shown in figure 3.4. Recalling our definition in the previous chapter, we equate the steady-state solution of equation (3.5), with the home range of the individual:

$$\underbrace{\nabla^2 u}_{\text{random motion}} - \underbrace{\beta \nabla \cdot (u\vec{\mathbf{x}})}_{\text{directed motion toward } \vec{\mathbf{x}}} = 0, \qquad (3.7)$$

where $\beta = c/d$. Since we are solving for the expected location of an individual, $u(x)$ is a probability density function, and integration of equation (3.7) over the region in which the individual moves must equal one. Thus, equation (3.7) is subject to the following constraint:

$$\int_{\Omega} u(x)dx = 1, \qquad (3.8)$$

where Ω is the region over which the individual is able to move. In this case, we assume the individual is moving in unbounded region that has no edges (i.e., $-\infty \le x \le \infty$).

3.4. SOLVING FOR PATTERNS OF SPACE USE

We first consider a simplified version of equation (3.5), in which the individual moves in a one-dimensional landscape. For convenience, and without loss of generality, we consider the case where the home range center is located at the origin $x = 0$. This yields the following equation:

$$\frac{\partial u}{\partial t} = \frac{\partial}{\partial x} (c \text{ sgn}(x)u) + d\frac{\partial^2 u}{\partial x^2}, \qquad (3.9)$$

where (sgn) is positive (+) for $x > 0$ and negative (−) for $x < 0$. Recalling our definition in the previous chapter, we now calculate the steady-state solution of equation (3.9), equating this with its home range:

$$d\frac{\partial^2 u}{\partial x^2} + \frac{\partial}{\partial x} (c \text{ sgn}(x)u) = 0. \qquad (3.10)$$

Integration of equation (3.10) and application of the constraint specified in equation (3.8) leads us to the following equation:

$$u(x) = \frac{\beta}{2} \exp(-\beta|x|). \qquad (3.11)$$

where $\beta = c/d$.

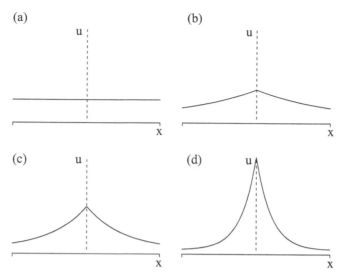

FIGURE 3.7. Home range patterns arising from localizing tendency model (eq. (3.7)). Panels show the expected pattern of space use in a single space dimension for different values of β: (a) $\beta = 0$, (b) $\beta = 1$, (c) $\beta = 2$, (d) $\beta = 5$. Solid lines show the probability density function $u(x)$ given by the solution of equation (3.11). Dashed lines indicate the location of the home range center.

Equation (3.11) tells us that, in one dimension, the pattern of space use that results from the localizing tendency model declines exponentially with distance from the home range center, and that the rate of decrease is governed by a single parameter β, which reflects the strength of directed movement towards the home range center relative to the strength of non-directed movement. Figure 3.7 shows solutions of equation (3.11) for different values of β. When β is zero, the individual has no bias in movement direction and its motion is purely random, giving rise to a uniform pattern of space use (figure 3.7a). When β is positive, the bias in movement directions toward its home range center gives rise to a home range in which space use declines monotonically as a function of distance from the home range center (figure 3.7b). As β increases, space use declines more rapidly with distance (figures 3.7c–d). A generalization of this approach for calculating territorial space use, which includes the possibility of nonlinear diffusive motion, can be found in White et al. (1996).

To calculate the expected pattern of space use for the red fox whose movements are shown in figure 3.4, we must solve the space use equation (3.7) for the biologically realistic case of movement in two dimensions. One option would be to solve this equation numerically; however, the radial symmetry in the underlying stochastic movement model allows us to simplify equation (3.7) and solve it analytically. Defining r as the distance from any point to the home range center, the advection and diffusion terms become

$$-\beta\nabla\cdot(u\vec{\mathbf{x}}) = \frac{\partial}{\partial r}(\beta u) \quad\text{and}\quad \nabla^2 u = \frac{1}{r}\frac{\partial^2}{\partial r^2}(ru)$$

respectively. Thus equation 3.7 becomes

$$\frac{\partial}{\partial r}(\beta u) + \frac{1}{r}\frac{\partial^2}{\partial r^2}(ru) = 0. \tag{3.12}$$

Expanding the second term in equation 3.12, we get

$$\frac{\partial^2 u}{\partial r^2} + \left(\frac{1}{r}+\beta\right)\frac{\partial u}{\partial r} = 0. \tag{3.13}$$

Using $r\exp(\beta r)$ as an integrating factor, we calculate that

$$\frac{\partial u}{\partial r} = A\frac{\exp(-\beta r)}{r}.$$

Since $u(r)$ is a probability density function, the area under $u(r)$ equals 1, thus the constant $A = -\beta^2/\pi$, so that

$$u(r) = \frac{\beta^2}{\pi}E_1(\beta r), \tag{3.14}$$

where E_1 is the exponential integral

$$E_1(u) = \int_u^\infty \frac{\exp(-w)}{w}\,dw. \tag{3.15}$$

3.5. PREDICTED RED FOX HOME RANGE

We can now use equation (3.14) to calculate the pattern of space use predicted by the localizing tendency model whose parameters we estimated from the red fox movements data shown in figure 3.4. The estimates of 0.085 km h^{-1} and 0.041 km^{2}h^{-1} for c and d calculated earlier imply that $\beta = c/d = 0.085/0.041 = 2.07$ km^{-1}. Substituting this value into equation (3.14) yields the predicted pattern of space use for the individual (figure 3.8).

We evaluate the prediction by comparing the predicted home range to a set of spatially independent relocations for the individual that provide an estimate of the individual's observed pattern of space use. We obtain the spatially independent relocations by resampling the high-frequency relocations shown in figure 3.4 at 6-hour intervals.

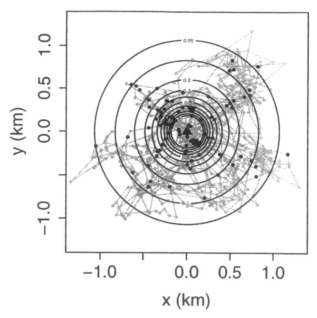

FIGURE 3.8. Ability of the localizing tendency home range model to predict the pattern of space use for the Cedar Creek red fox. Gray lines indicate the distribution of high-frequency (10 minute) relocations shown in figure 3.4. The points in the figure (•) are spatially independent relocations obtained by resampling the high-frequency relocations at 6-hour intervals. The location of the individual's home range center is also shown (▲). Contour lines indicate the probability density function for the individual's home range predicted by the localizing tendency model. The value of the model's parameter $\beta = 2.07$ was calculated from the measurements of fine-scale movement behavior shown in figure 3.6.

Visual inspection of the expected pattern of space use shows that the localizing tendency model predicts a radically uniform pattern of space use for the individual around the home range center (figure 3.8). Overlaid on the figure are the spatial distribution of the high-frequency relocations (gray lines) and a spatially independent set of relocations for the individual (•), obtained by resampling the relocations depicted in figure 3.4 at 6-hour intervals. The spatial distribution of high-frequency relocations imply a less radially uniform pattern of space use for the individual (eq. (3.14)), but the relatively small number of spatially independent relocations in this dataset are insufficient to adequately evaluate the pattern of space use predicted by the localizing tendency model.

To further evaluate the ability of the localizing tendency model to characterize carnivore home range patterns, we turn to an alternative dataset composed of a large number of spatially independent relocations collected as part of a study of coyote (*Canis latrans*) home ranges in south-central Washington.

3.6. COYOTE HOME RANGE PATTERNS

Color plate 1 shows observations of coyote home ranges collected at Hanford
Arid Lands Ecological Reserve (ALE) in south-central Washington by Crabtree
(1989). The dataset comprises radio locations of known individuals belonging
to six contiguous packs in the western portion of the ALE. Weekly relocations of the individuals were collected over an eighteen-month period between
December 1985 and June 1987 from high-accuracy triangulations ($0.5°$ error),
recorded by two fixed monitoring stations that had "line-of-sight" coverage
over the ~ 150 km^2 study area; see Crabtree (1989) for details.

For the red fox at Cedar Creek, we were able to use the high-frequency
relocation data (figure 3.8) to estimate the coefficients of the underlying redistribution kernel of the localizing tendency model (eq. (3.3)) and use this to
calculate an independent estimate for the value of β in equation (3.14). We then
were able to use this equation to predict the pattern of space use that results
from the red fox's underlying pattern of fine-scale movement.

The Hanford data (color plate 1), however, consists of low-frequency, weekly
relocations over an eighteen-month period that are not suitable for calculating
the coefficients of the underlying redistribution kernel (eq. (3.3)) for the coyote
at Hanford. Instead we adopt a different approach. Rather than calculating the
parameters of the underlying redistribution kernel to obtain an independent
estimate for β and then using this to predict space use, we fit the expected
steady-state pattern of space use that arises from the localizing tendency model
equation (3.7) to relocations that are sufficiently separated in time that they
constitute spatially independent weekly relocations. In doing so, we are no
longer producing an *a priori* prediction for the pattern of space use, but rather
estimating the value of β that best matches the fit of the model to the observed
distribution of relocations.

We fit the model (eq. (3.7)) to the relocation data for the centrally
located Hopsage pack (red points in color plate 1) using maximum likelihood.
Equation (3.7) contains a single parameter β whose value we estimate by using
a log-likelihood function $l(\beta)$ to measure goodness-of-fit:

$$l(\beta) = \sum_{j=1}^{n_r} \ln u(x_j, y_j), \qquad (3.16)$$

where $u(x_j, y_j)$ is the height of the probability density function at point (x_j, y_j),
given by the steady-state solution from equation (3.7), and where (x_j, y_j) are
the coordinates of the jth relocation, $j = 1 \ldots n_r$, where n_r is the total number
of relocations for the Hopsage pack. As at Cedar Creek, we used the centroid
of the relocation observations as the location of the home center $\vec{\mathbf{x}} = [x_H, y_H]$.

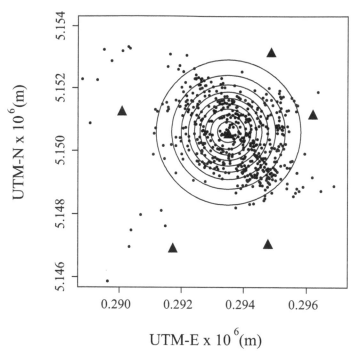

FIGURE 3.9. Contour lines showing the probability density function $u(x, y)$ for the home range of the central Hopsage coyote pack at Hanford ALE obtained by fitting the Holgate-Okubo localizing tendency model (eq. (3.7)) to relocation data (•) collected by Crabtree (1989) ($n_{reloc} = 480$ relocations). The contour interval is 2, in density units scaled so that both the domain area A and integral of $u(x, y)$ equal 1. Home range center of the Hopsage pack and neighboring packs, as estimated by the centroids of the relocation data for each pack, are shown (▲). The fitting procedure is described in the text, and the maximum likelihood estimate for the ratio of directed to non-directed movement β, from maximizing equation (3.16), is 12.34. Positions are indicated in Universal Transmercator (UTM) coordinates (east, north).

The model was fitted to the data by maximizing equation (3.16) with respect to β. More details on the model-fitting procedure can be found in appendix F.

The localizing tendency model gives a relatively poor fit to the observed pattern of relocations of the Hopsage coyote pack (figure 3.9). The pattern of space use predicted reflects the movement rules underlying equation (3.7), in which the only aspect of movement that varies spatially is the bias in movement direction that is always directed toward the home range center. As a result, the localizing tendency model gives rise to a radially symmetric home range that while mathematically elegant, is not particularly realistic biologically.

One potential explanation for the poor fit seen in figure 3.9 is that spatial heterogeneity in resource availability, habitat, or topography generates spatial variation in the movement behavior of the individuals in different parts of the landscape, giving rise to a non–radially uniform pattern of space use by the pack. We will be examining this issue later in chapter 7. However, at least in the case of the coyote data at Hanford, this explanation seems unlikely given the relatively uniform sagebrush environment in eastern Washington (see figure 1a and color plate 10).

An alternative explanation is that neighboring packs may be influencing the movement behavior of the Hopsage individuals. As can be seen in figure 3.9, several neighboring packs occupy the areas around the Hopsage pack. The importance of neighbors is highlighted when we fit the home range model to the surrounding coyote packs in the Hanford region (color plate 2). As can be seen from the figure, the model does not capture the spatial pattern of home ranges across the region. In particular, the model does a poor job of characterizing the boundaries between adjacent groups, suggesting a high degree of home range overlap that is not present in the data. We address this issue in the following chapter, developing a mechanistic home range model that accounts for interactions with neighboring conspecifics.

3.7. SUMMARY

In this chapter, we derived a simple mechanistic home range model, the Holgate-Okubo localizing tendency model, from an underlying set of movement rules describing the fine-scale movement behavior of individuals. The starting point was a stochastic movement model in which individuals preferentially move in the direction of a home range center corresponding to a densite or core foraging area. Using a dataset of high-frequency relocations for a red fox, we showed how the coefficients of the underlying movement model can be estimated from observations of the individual's fine-scale movement behavior. Using the approaches described in chapter 2, we then derived the predicted pattern of space use that results from the localizing tendency redistribution kernel. After a brief mathematical analysis of the model's predictions in a single space dimension, we showed how the model's prediction for space use in two dimensions compared to the fox's observed pattern of space use. We then examined the ability of the localizing tendency model to characterize the patterns of space use by coyotes in south-central Washington. The fit of the model to a dataset of coyote relocations highlighted two shortcomings that arise from the static nature of the movement rules underlying the localizing tendency model. The first is the lack of any influence of exogenous heterogeneities such as spatial variation in resource availability, habitat, or topography on movement behavior of individuals. The second is that, in the localizing tendency model, the movements of individuals are unaffected by the presence of neighboring individuals. This latter issue is explored in more detail the following chapter.

CHAPTER FOUR

A Model Based on Conspecific Avoidance

In the localizing tendency home range model developed in chapter 3, individuals exhibited a fixed bias in their fine-scale movement behavior toward a home range center. However, the model predicted a circular home range, which gave a relatively poor fit to a dataset of coyote relocations. In particular, the model was unable to capture the relatively sharp boundaries between neighboring home ranges (see color plate 2). In this chapter we develop an alternative mechanistic home range model in which individuals exhibit an avoidance response to the scent marks of conspecifics. Our analysis closely follows that of Moorcroft et al. 1999.

Conspecific avoidance is a widespread phenomenon in many carnivore species. Sometimes it arises as a result of direct aggressive encounters between individuals, but more usually it occurs indirectly, as a result of interactions through proximate signaling cues (MacDonald 1980a). In foxes, coyotes, wolves, and many other carnivores, the dominant proximate signaling cue underlying conspecific avoidance is scent marking. Scent marking serves a number of functions in mammalian populations (Ralls 1971), but in carnivores it is closely linked to the formation and maintenance of a home range (Ralls 1971; Eisenberg and Kleiman 1972; Johnson 1973; Rothman and Mech 1979; Barrette and Messier 1980; MacDonald 1980a; Gese and Ruff 1996). The high marking rates of individuals and the persistence of scent marks in the environment means that carnivore home ranges typically contain large numbers of scent marks that individuals continually encounter during movement. For example, Mills estimated that brown hyena (*Hyaena brunnea*) home ranges in the Kalahari contained 20,000 active scent marks and individuals encountered scent marks every few minutes (Gorman and Mills 1984). Estimates for coyotes and wolves suggest similar high rates of scent-mark interaction (Peters and Mech 1975; Bowen and McTaggart-Cowan 1980).

There are numerous anecdotal accounts of intruders retreating following encounters with foreign scent marks (for example, Peters and Mech (1975) and Bowen and McTaggart-Cowan (1980)). More direct evidence of scent marks influencing the movement behavior of individuals and altering patterns of space use comes from a study by MacDonald (1980a), which showed that the spatial extent of a red fox's home range was closely correlated with the spatial extent of its scent marking, with the individual turning back at the edges of its area as a result of encounters with foreign scent marks. Moreover, MacDonald showed that it was possible to manipulate the spatial extent of the individual's movements by altering the locations of foreign scent marks at the edge of the individual's home range.

In addition to affecting movement, scent marks also influence the scent-marking behavior of many carnivores. The phenomenon of "over-marking" is well known to dog owners and canid biologists alike and hardly requires further description. However, a number of studies have documented this behavior in natural canid populations. For example, Peters and Mech (1975) showed that wolf scent-marking rates are significantly elevated in areas where the concentration of active scent marks is higher. Similar responses have been documented in many canids, felids, and other carnivores (Wells and Bekoff 1981).

In summary, field observations suggest that the two following "behavioral rules" apply for the movement and scent-marking behavior of many carnivore species:

1. Foreign scent marks influence an individual's movement behavior, increasing the likelihood of movement toward the interior of its home range.
2. Encounters with foreign scent marks increase an individual's subsequent rate of scent marking.

We now formulate a mechanistic home range model that reflects these observations.

4.1. MODEL FORMULATION

Initially, we focus on a pairwise interaction between individuals belonging to two groups, U and V, that are of equal size. We begin by considering the effect of foreign scent marks on the movement behavior of the individuals.

Individual Movement Behavior

As in the previous chapter, we start by specifying a redistribution kernel describing the probability of moving from point $\mathbf{x}'$ to point $\mathbf{x}$ in a time step of length τ, but in this case, the turning kernel for the individual reflects observation (1) above, namely, that individuals exhibit an avoidance response to foreign scent marks. We do this by considering a modification of equation (3.3) in which the distribution of movement directions K_τ depends not only on the direction to the home range center $\widehat{\phi} = \tan^{-1}(y - y_H/x - x_H)$, but also on the level of foreign scent mark encountered.[1] Using notation similar to that in the previous chapter, the redistribution kernel for individuals of pack U can be expressed as

$$k^u(\mathbf{x}', \mathbf{x}, \tau, t) = \underbrace{\frac{1}{\rho} f_\tau(\rho) K_\tau^u(\phi, \widehat{\phi})}_{\text{probability density for moving from } \mathbf{x}' \text{ to } \mathbf{x}}, \tag{4.1}$$

where K_τ^u is the von Mises distribution (eq. (3.2)) with the concentration parameter κ_τ^u dependent upon $q(\mathbf{x}, t)$, the density of scent marks of pack V.

We can write an analogous equation for the redistribution kernel of individuals of the V pack:

$$k^v(\mathbf{x}', \mathbf{x}, \tau, t) = \frac{1}{\rho} f_\tau(\rho) K_\tau^v(\phi, \widehat{\phi}), \tag{4.2}$$

where K_τ^v is the von Mises distribution (eq. (3.2)) with the concentration parameter κ_τ^v dependent upon $p(\mathbf{x}, t)$, the density of scent marks of pack U.

We incorporate the avoidance response by making the degree of bias in movement direction toward the individual's home range center a linearly increasing function of the foreign scent-mark density. For example, if, as in the previous chapter, we use a von Mises distribution (eq. (3.2)) for the turning kernel of the individuals, we can make the concentration parameters of pack U and pack V turning kernels, κ_τ^u and κ_τ^v, increase in proportion to functions of the density of foreign scent marks encountered:

$$\kappa_\tau^u = b\rho_\tau q(\mathbf{x}, t), \tag{4.3}$$

$$\kappa_\tau^v = b\bar{\rho}_\tau p(\mathbf{x}, t). \tag{4.4}$$

Here q and p are the scent-mark densities of pack U and pack V respectively, $\bar{\rho}_\tau$ is the mean step length over a time step of length τ, and the parameter b governs the sensitivity of the individual's distribution of movement directions to foreign scent marks. The parameter b is similar to the corresponding parameter in the Holgate-Okubo localizing tendency model (see eq. (3.4)); however, in this case, it is the bias per unit distance moved per unit density of scent marks encountered.

Figure 3.2 shows how the shape of the individual's distribution of movement directions changes as a function of increasing the concentration parameter. In the conspecific avoidance model formulation, this corresponds to either an

---

[1] Alternatively we could assume that the distribution of possible distances traveled in a time step of length τ, $\rho(\tau) = |\mathbf{x}' - \mathbf{x}|$, depends upon foreign scent-mark levels.

increasing sensitivity in the individual's movement direction to foreign scent marks (b), or, for a given sensitivity b, an increase in the density of foreign scent marks encountered ($q(\mathbf{x}, t)$ or $p(\mathbf{x}, t)$, depending on whether the individual belongs to pack U or pack V).

Substituting equations (3.2), (4.3), and (4.4) into equations (4.1) and (4.2) yields expressions for the redistribution kernels for the two packs. Substituting these into equation (2.16) and using the methods described in appendix E to simplify the resulting equations, we obtain the following equations for the expected patterns of space use by the two groups:

$$\frac{\partial u}{\partial t} = dV^2 u \quad - \quad cV \cdot \left[u\vec{\mathbf{x}}_u q \right] \tag{4.5}$$

$$\frac{\partial v}{\partial t} = \underbrace{dV^2 v}_{\text{random motion}} \quad - \quad \underbrace{cV \cdot \left[v\vec{\mathbf{x}}_v p \right]}_{\text{directed motion}} \tag{4.6}$$

where $\vec{\mathbf{x}}_u$ and $\vec{\mathbf{x}}_v$ are unit vectors pointing toward the den sites for pack U and V, and the advection coefficient c and diffusion coefficient d are given by

$$c = \lim_{\tau \to 0} \frac{b\bar{\rho}_\tau^2}{2\tau} \quad \text{and} \quad d = \lim_{\tau \to 0} \frac{\bar{\rho}_\tau^2}{4\tau}. \tag{4.7}$$

Comparing equations (4.5) and (4.6) to equation (3.5) we see that the magnitude of the directed movement term c for the two packs varies with the density of foreign scent marks encountered ($q(\mathbf{x}, t)$ and $p(\mathbf{x}, t)$). Thus, the patterns of space use by the U and V packs depends on the spatial distribution of scent marks by the two groups.

Scent-Marking Equations

To complete the model, we must now formulate equations expressing how the spatial distribution of scent marks changes as a result of scent marking by individuals in the two packs. Suppose that in the absence of foreign scent marks individuals scent mark at a rate l, and that the marks decay as a result of aging at rate μ. Suppose further that consistent with observation (2) above, the presence of foreign scent marks causes individuals to increase their marking rate by an amount proportional to the local density of foreign scent marks encountered. Recalling that $p(\mathbf{x}, t)$ and $q(\mathbf{x}, t)$ are the expected density of scent marks for packs U and V, we can write equations that describe the rate of change of

p and q at each point in space $\mathbf{x} = (x, y)$:

$$\frac{\partial p}{\partial t}(\mathbf{x}, t) = Nu(\mathbf{x}, t)(l + mq(\mathbf{x}, t)) - \mu p(\mathbf{x}, t), \qquad (4.8)$$

$$\frac{\partial q}{\partial t}(\mathbf{x}, t) = \underbrace{Nv(\mathbf{x}, t)(l + mp(\mathbf{x}, t))}_{\text{scent-mark deposition}} - \underbrace{\mu q(\mathbf{x}, t)}_{\text{scent-mark decay}}, \qquad (4.9)$$

where m denotes the sensitivity of the marking response to foreign scent marks and N is the number of individuals in each pack.

Boundary Conditions

The space use equations (4.5) and (4.6) have associated boundary conditions describing the behavior of the solutions at the boundary of the region under consideration. As we noted in section 2.4, a suitable choice of boundary conditions for equations (4.5) and (4.6) are "zero flux" boundary conditions, indicating that movements and interactions remain in a finite, self-contained region corresponding to the study area. These boundary conditions guarantee that the total number of individuals in each pack remain constant over time.

4.2. EQUATIONS FOR SPACE USE

Before analyzing the home range model, it is convenient to reduce the number of parameters by nondimensionalization. An excellent general introduction to the technique of nondimensionalization can be found in Segel (1972). Equations (4.5)–(4.6) and (4.8)–(4.9) can be nondimensionalized by introducing the following variables:

$$x^* = \frac{x}{L}, \quad y^* = \frac{y}{L}, \quad t^* = t\mu, \quad u^* = L^2 u,$$

$$v^* = L^2 v, \quad p^* = \frac{L^2 \mu p}{Nl}, \quad q^* = \frac{L^2 \mu q}{Nl} \qquad (4.10)$$

$$d^* = \frac{d}{\mu L^2}, \quad c^* = \frac{clN}{\mu^2 L^3}, \quad m^* = \frac{mN}{\mu L^2},$$

where L is a characteristic length scale that is related to the area A ($L = A^{\frac{1}{2}}$) of the domain Ω over which the equations are to be solved (the study area). After nondimensionalization, the time scale for the system is now the rate of decay of scent marks μ, the domain area is now equal to 1, and the scent-mark densities p^* and q^* are now scaled to the background rate of marking and the rate of scent-mark decay.

Making the above substitutions into equations (4.5)–(4.6) and (4.8)–(4.9), and then dropping the asterisks for notational simplicity, gives us

$$\frac{\partial u}{\partial t} = d\nabla^2 u - c\nabla \cdot \left[u \vec{\mathbf{x}}_u q \right],$$ (4.11)

$$\frac{\partial v}{\partial t} = d\nabla^2 v - c\nabla \cdot \left[v \vec{\mathbf{x}}_v p \right],$$ (4.12)

$$\frac{dp}{dt} = u(1 + mq) - p,$$ (4.13)

$$\frac{dq}{dt} = v(1 + mp) - q.$$ (4.14)

As in chapter 3, we determine the home ranges by calculating the time-independent solutions of the above equations. Applying a steady-state condition to equations (4.11)–(4.14) yields the following system of equations:

$$0 = \nabla^2 u - \beta \nabla \cdot \left[u \vec{\mathbf{x}}_u q \right],$$ (4.15)

$$0 = \nabla^2 v - \beta \nabla \cdot \left[v \vec{\mathbf{x}}_v p \right],$$ (4.16)

$$0 = u(1 + mq) - p,$$ (4.17)

$$0 = v(1 + mp) - q.$$ (4.18)

In our nondimensionalized steady-state model there are only two parameters: the sensitivity of an individual's scent-marking rate to encounters with foreign scent marks m, and the relative importance of directed versus random movement

$$\beta = \frac{c}{d}.$$ (4.19)

The parameter β is similar to the one that appeared in the nondimensionalized form of the localizing tendency model developed in chapter 3. However, in the case of the conspecific avoidance model, β is a measure of the individual's directed motion per unit of scent-mark density encountered. Its value reflects the following combination of dimensional parameters: $\frac{c}{d} \frac{l}{\mu L}$. The nondimensionalized parameter m is given by the following combination of dimensional parameters: $\frac{m}{\mu} \frac{N}{L^2}$.

4.3. EMPIRICAL EVALUATION OF THE MODEL

We now investigate the ability of the conspecific avoidance model to account for the patterns of coyote space use seen in color plate 1, comparing the fit of the model to the fit obtained from the Holgate-Okubo localizing tendency model (eq. (3.7)). In order to do this, we must first generalize the above model from a pairwise interaction between two groups to the case of multiple interacting packs.

Equations for Interacting Packs

Equations (4.15)–(4.18) readily generalize to the case of multiple interacting packs. The home ranges for n packs $i = 1 \ldots n$ are described by the solution to

$$0 = \nabla^2 u^{(i)} - \beta \nabla \cdot \left[u^{(i)} \vec{\mathbf{x}}_i \sum_{j \neq i}^{n} p^{(j)} \right], \qquad i = 1 \ldots n, \qquad (4.20)$$

where $u^{(i)}(\mathbf{x}), (i = 1 \ldots n)$, are probability density functions describing the home ranges of the n groups. Note that in equation (4.20), the strength of directed movement for individuals within each of the n packs is determined by the local density of foreign scent marks ($\sum p^{(j)}, j \neq i$).

The above equations for space use are coupled to a corresponding system of algebraic equations describing the steady-state distribution of scent marks $p^{(i)}(\mathbf{x})$ for each pack at each point in space within the study region:

$$0 = u^{(i)} \left[1 + m \sum_{j \neq i}^{n} p^{(j)} \right] - p^{(i)}, \qquad i = 1 \ldots n. \qquad (4.21)$$

Here the explicit spatial dependencies of u and p have been dropped for notational convenience.

The space use equations (4.20) have associated boundary conditions which describe the behavior of the solutions at the boundary $\partial\Omega$ of the domain Ω. As we noted earlier, a suitable choice of boundary conditions are "zero flux" boundary conditions, indicating movement and interaction in a finite, self-contained region corresponding to the study region. Mathematically, these are given by

$$\left[\nabla u^{(i)} - \beta u^{(i)} \vec{\mathbf{x}}_i \sum_{j \neq i}^{n} p^{(j)} \right] \cdot \vec{\mathbf{n}} = 0, \qquad (4.22)$$

where $\vec{\mathbf{n}}$ is the outwardly oriented unit vector normal to the edge of the domain $\partial\Omega$.

Data and Model Fitting

As in chapter 3, we fit the model to the coyote relocation data collected at Hanford ALE (see color plate 1), using a log-likelihood function to measure goodness-of-fit:

$$l(\beta, m) = \sum_{i=1}^{n} \sum_{j=1}^{n_{reloc}^{(i)}} \ln u^{(i)}(x_{ij}, y_{ij}). \qquad (4.23)$$

Here $u^{(i)}(x_{ij}, y_{ij})$ is the height of the probability density function (pdf) for expected space use by pack i at point (x_{ij}, y_{ij}), given by the steady-state solution of equations (4.20)–(4.21).

Since the home range model equations (4.20)–(4.21) cannot be solved analytically for the biologically realistic case of home ranges in two-dimensional space, for each set of values of β and m, equations (4.20)–(4.21) are solved numerically. We simulate the model equations on a 12.5 km × 11.0 km domain that encompasses the relocation dataset, discretizing the equations on a spatial grid at a resolution of 100 m × 100 m, since fine-scale approximations of the spatial derivatives are necessary for accurate simulation of the model equations (see appendix G for more details on the numerical simulation procedure).

In the first analysis, we fit the conspecific avoidance model to relocation data for the centrally located Hopsage pack (see color plate 1). In the second analysis, we use the model to characterize the regional home range patterns at Hanford ALE, fitting the model to the relocation data for six contiguous packs in the region.

Characterization of a Single Home Range

Visual inspection of the home range fit for the Hopsage pack shows that the conspecific avoidance model matches the spatial distribution of relocations more closely than does the localizing tendency model (figure 4.1). In contrast to equation (3.7), which predicted a circular home range, the shape of the contour lines around the home range center predicted by equations (4.20)–(4.21) indicate that the shape of the probability density function for expected space use is being influenced by the presence of neighboring home ranges.

The goodness-of-fit of the conspecific avoidance model can be tested against the fit of the localizing tendency model by examining the log-likelihood scores for the two model fits. Since the models are not nested, an appropriate metric for comparing the models is the Akaike information criterion (AIC), which states that the best model is the one with the lowest AIC score = $-2 \times$ (log-likelihood) $+ 2 \times$ (number of model parameters) (Burnham and Anderson 2002). While the AIC can be derived directly from information theory (Burnham and Anderson 2002), an intuitive explanation for the AIC is that it assesses the goodness-of-fit of a model taking into account its complexity, penalizing models that have more parameters (Hilborn and Mangel 1997). A good rule of thumb is that models whose AIC scores differ by more than 10 (i.e., ΔAIC > 10) are substantially different. Examination of the AIC scores for the two models (table 4.1) shows that switching from a constant bias in movement direction (eq. (3.7)) to directional movement induced by encounters with foreign scent marks (eqs. (4.20)–(4.21)) reduces the AIC from -1481.8

TABLE 4.1. Details of the home range model fits (figures 4.1, and color plate 3 to relocation data collected at Hanford ALE. Parameter values, log-likelihood scores ($l(\theta)$ where
θ is the set of model parameters), and Akaike Information Criterion (AIC) scores are
given for the fit of the conspecific avoidance home range model (eqs. (4.20)–(4.21)) and
for the fit of the Holgate-Okubo localizing tendency model (eq. (3.7)) shown in figure 3.9
of the previous chapter. The single home range model fits used relocations collected on
individuals belonging to the Hopsage pack. In the regional fit, the conspecific avoidance home range model (eqs. (4.20)–(4.21)) was fitted to relocation data for the six
contiguous pack home ranges in Hanford ALE study area. n_{total} is the total number of
data points used in each model fit $n_{total} = \sum_{i=1}^{n} n_{reloc}^{(i)}$ (see eq. (4.23)). Note that in
equation (3.7), β indicates the ratio of directed movement to non-directed movement,
while in equations (4.20)–(4.21), β indicates the ratio of directed movement per unit of
scent-mark density encountered relative to the strength of non-directed movement.

	Single Home Range		Regional Fit
	Conspecific avoidance	Holgate- Okubo	Conspecific avoidance
# Packs (n)	1	1	6
Equations	4.20–4.21	3.7	4.20–4.21
Parameter			
β	5.618	12.300	0.5467
m	0.083	–	0.131
$l(\theta)$	801.3	741.9	3106
AIC	−1598.6	−1481.8	−6208
	$n_{total} = 480$		$n_{total} = 2325$

to −1598.6 (ΔAIC = 116.8), implying that the scent-mark avoidance model
provides a substantially better explanation for observations than the localizing
tendency model.

Regional Home Range Patterns

Fits of the conspecific avoidance home range model (eqs. (4.20)–(4.21)) to the
relocation data for the six contiguous packs at Hanford are shown in figure 4.2
and color plate 3. Compared to the single pack fit (figure 4.1), the regional fit
of the model to the centrally located Hopsage pack suggests a more exclusive
pattern of space use (figure 4.2). Including the relocations from the surrounding
groups into the model fits increases the degree of platykurtosis in the probability
density functions for expected space use, indicating a more uniform intensity of
space use within sharply defined home range boundaries beyond which space
use declines sharply (color plate 3).

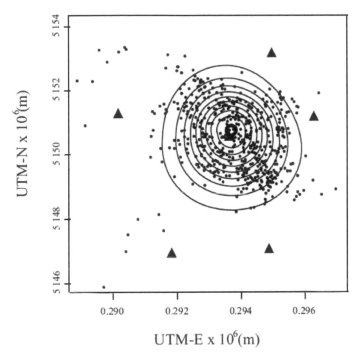

FIGURE 4.1. Contour lines showing the probability density function $u(x, y)$ for the home range of the central Hopsage coyote pack at Hanford ALE obtained by fitting the conspecific avoidance model (eqs. (4.20) and (4.21)) to relocation data (●) collected by Crabtree (1989). The contour interval is 2, in density units scaled so that both the domain area A and the integral of $u(x, y)$ are equal to 1. Home range centers of the Hopsage pack and neighboring packs, as estimated by the centroids of the relocation data for each pack, are shown (▲). The fitting procedure is described in the text, and the maximum likelihood values and parameter estimates for β and m are given in table 4.1. Pack positions are indicated in Universal Transmercator (UTM) coordinates (east, north).

Through the influence of conspecific scent marks on individual movement, the conspecific avoidance home range model captures the spatial pattern of home ranges across the Hanford region well, capturing the location of the boundaries between adjacent home ranges (color plate 3). The degree of overlap between neighboring packs appears to vary: the central Hopsage pack home range has sharp boundaries in the NE section of the study area where the neighboring home range centers are in close proximity, while the boundaries between the other, more distant neighbors are more overlapping (color plate 3). At the periphery of the region, the ability of the model to describe observed patterns of space use is more limited. The shape of the home range around these exterior boundaries is determined by the "zero-flux" boundary conditions at the edges of the domain (eq. (4.22)), rather than as a result of interactions with foreign scent marks of adjacent packs (color plate 3).

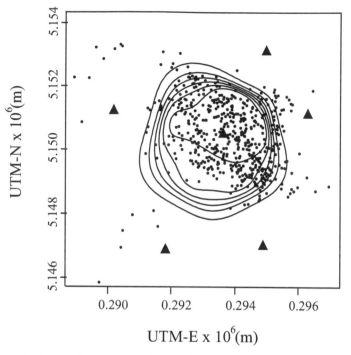

UTM-E x 10^6(m)

FIGURE 4.2. Contour lines showing the probability density function $u(x, y)$ for the home range of the centrally located Hopsage coyote pack obtained by fitting the conspecific avoidance home range model (eqs. (4.20)–(4.21)) to relocation data (•) for all six coyote groups at Hanford ALE collected by Crabtree (1989). As in figure 4.1, the contour interval for $u(x, y)$ is 2, and the home range centers for each pack are shown (▲). Maximum likelihood values and estimates for β and m are given in table 4.1.

Model Predictions

In this section we use the results of the regional fit of the conspecific avoidance model to predict (i) the expected spatial distribution of scent marks across the region, (ii) spatial variability pattern in individual movement behavior, and (iii) the effects of pack removal upon home range patterns within the region.

Patterns of Scent Marking

Since the spatial pattern of home ranges arises from interactions between individuals and scent marks, the fit of the conspecific avoidance model yields information about the expected distribution of scent marks across the region (figure 4.3). The results of the regional fit at Hanford (color plate 3) suggest that in the interior of the domain, higher concentrations of scent marks should occur along the boundaries between neighboring home ranges. In particular, in

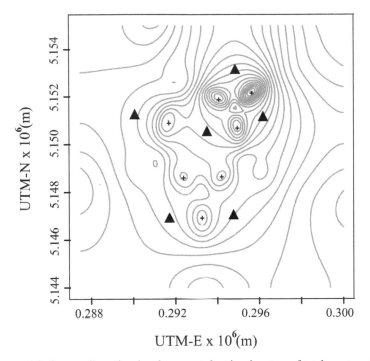

FIGURE 4.3. Contour lines showing the expected regional pattern of total scent-mark density $\sum_{i=1}^{n} p^{(i)}$ within the Hanford ALE study area, obtained from the fit of the conspecific avoidance model (eqs. (4.20)–(4.21)) shown in color plate 3. Peaks in scent-mark density are indicated (+). Scent-mark densities across the domain vary between 1.1 and 22.1, in rescaled units (eq. (4.21)), shown using a contour interval of 1.

the NE section of the domain where three home range centers lie in close proximity, high scent-mark concentrations are expected (figure 4.3). The absence of any high concentrations at the periphery of the region is due to the absence of interacting packs along these home range boundaries, which, as noted in section 4.1, are governed by the boundary conditions.

Patterns of Fine-Scale Movement

The mechanistic nature of the home range model used in this analysis means that the parameters (table 4.1) reflect the underlying movement behavior of individuals that can be verified against measurements obtained on the fine-scale movement and scent-marking behavior of individuals. For example, we can use the model fit to calculate the predicted distribution of animal movement directions in different parts of the study area. Replacing $q(\mathbf{x}, t)$ with $\sum_{j \neq i}^{n} p^{(j)}(\mathbf{x}, t)$ in equation (4.3), then inserting into equation (3.2), we obtain the following equation for the predicted distribution of movement directions for individuals in each pack i $(i = 1 \ldots n)$ as a function of their spatial position:

$$K(\phi) = \frac{\exp\left[b\bar{\rho}_\tau \sum_{j \neq i}^{n} p^{(j)}(\mathbf{x}, t) \cos(\phi - \widehat{\phi})\right]}{2\pi I_0 \left(b\bar{\rho}_\tau \sum_{j \neq i}^{n} p^{(j)}(\mathbf{x}, t)\right)}$$

where $\widehat{\phi}$ is the direction of the home range center, $\bar{\rho}_\tau$ is the mean distance between successive relocations, and b is the sensitivity of movement direction to foreign scent-mark density. The model fit provides the values of the scent-mark density at each location for each pack $p^{(i)}(\mathbf{x}, t)$ (see eq. (4.21) and figure 4.3). The value of b is related to the bias parameter β, whose value is estimated as part of the model-fitting procedure (see table 4.1). If, as is commonly observed, distribution of move lengths between relocations is approximately exponential, we can relate the first and second moments $\bar{\rho}_\tau^2 = 2\bar{\rho}_\tau^2$, and thus $b = \beta$ (see eqs. (4.3), (4.7), and (4.19)). Resolving the density of scent marks "foreign" to a particular pack $\sum_{j \neq i}^{n} p^{(j)}(x, y)$ (figure 4.4a), then substituting into equation (4.24), along with the estimate of $b = \beta$, (table 4.1), yields predictions for the direction and magnitude of bias in movement direction across the domain, expressed as distributions of expected movement direction in different regions of the study area. For example, figure 4.5b shows the predicted distribution of movement directions of the members of the Hopsage pack at two different locations in the study region.

Effects of Pack Removal

The underlying formulation of the conspecific avoidance model, of movement in response to the scent marks of neighboring packs, results in home ranges whose size and shape are influenced by the location of adjoining home ranges. This effect, particularly apparent in the regional home range fit at Hanford (color plate 3), contrasts with the biologically unrealistic property of the localizing tendency model developed in the previous chapter, where the movement behavior of individuals is completely uninfluenced by the presence of neighboring home ranges.

The mutual interdependency in the size and shape of adjoining home ranges captured by this formulation enables the model fits to be used to predict the consequences of pack introduction or removal (color plate 4). For example, color plate 4 shows the pattern of space use that arises following the removal of the centrally located Hopsage pack. Prior to their removal, the Hopsage pack at Hanford has almost exclusive use of the center of the domain (color plate 3). However, following the removal of the Hanford pack, neighboring packs expand their home ranges into the unoccupied central area (color plate 4). In addition, due to the reduction in the density of packs in the region, the home ranges of the remaining packs become more overlapping (color plate 4). Accompanying

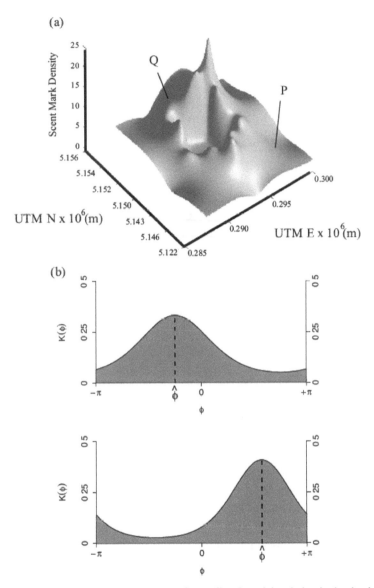

FIGURE 4.4. (a) Surface height indicates the predicted spatial variation in the density of foreign scent marks $\sum_{j \neq i}^{n} p^{(j)}$ encountered by members of the Hopsage pack in the Hanford ALE study region. (b) Predicted distribution of turning angles $K(\phi)$ in the regions around points P and Q labeled in (a). ϕ indicates movement direction relative to UTM grid north. Around P, the turning distribution is relatively uniform, indicating movement is relatively non-directional (isotropic). Around Q the turning distribution is less uniform, indicating movement is more directional (nonisotropic). $K(\phi)$ was calculated using the relationship between the density of foreign scent marks $\sum_{j \neq i}^{n} p^{(j)}$ and distribution of turning angles (eq. (4.24)), assuming a mean distance moved between successive relocations $\bar{\rho}_\tau$ of 0.5 km, and using a value of 0.547 for b obtained from the fits of equations (4.15)–(4.18) ($b = \beta$, see eqs. (4.3), (4.7), and (4.19) and table 4.1).

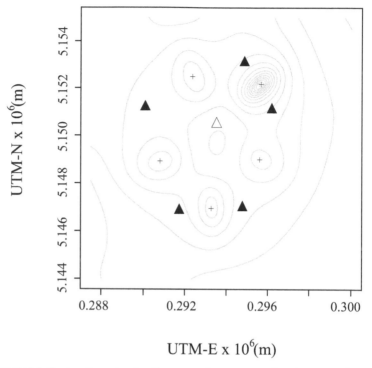

FIGURE 4.5. Contour lines showing the expected new pattern of total scent-mark density following the removal of the Hopsage pack shown in color plate 4. The pattern of scent-mark density prior to removal is shown in figure 4.3. Peaks in scent-mark density are indicated (+). Scent-mark densities across the domain vary between 0.9 and 15.1, in rescaled units (eq. (4.21)), shown using a contour interval of 1.

these changes in patterns of space use are changes in the spatial distribution of scent marks (figure 4.5). Following the removal, the peaks in scent-mark density present between the Hopsage pack and each of its neighbors (figure 4.3) disappear, while the peaks in scent-mark density between the remaining packs alter in shape and size, as patterns of space use by the remaining groups shift in response to the removal (figure 4.5). This is particularly evident in the NE region, where the three ridges of high scent-mark density present prior to the removal (figure 4.3) are replaced by a single ridge between two remaining packs (figure 4.5). Similarly, if the location of the denning area of a newly introduced pack can be predicted or is observed, the model fits can be used to predict the new arrangement of home ranges and patterns of scent marks.

The results of the model fitting provides empirical support for the behavioral rules underlying the conspecific avoidance home range model. In particular, the results of the likelihood analysis support for this formulation in which foreign scent-marks influence the movement behavior of individuals (eq. (4.20)). This provides a better fit to observed relocation patterns than the localizing tendency model (eq. (3.7)), in which individuals exhibit a constant bias in movement direction (table 4.1).

The shape of the probability density functions for the regional fit at Hanford suggests uniform space use by packs in the home range interior and well-defined home range boundaries. This shape implies relatively exclusive territories, differing markedly from the typical distributions obtained using statistical home range models. For example, the fit of a conventional bivariate normal statistical home range model to the Hanford home range data (color plate 5) exhibits two of the classic problems of the bivariate normal model: the home ranges are inappropriately peaked, and they have unrealistically long tails. Comparing the regional fit for the six contiguous packs (figure 4.2 and color plate 3) to the fit of a single group (figure 4.1) suggests that simultaneously fitting the model to relocation data collected on adjacent groups improves the characterization of home range boundaries, capturing variation in the degree of exclusivity in space use along different edges of the home range.

The ability of the conspecific avoidance model to accurately represent home range patterns at the periphery of the study areas was limited due to the influence of the boundary conditions. As we later show in chapter 7, if home ranges are located in a restricted area such as a steep-sided valley, the influence of terrain on the movement behavior of individuals can be incorporated into the underlying movement model. In the case of Hanford, where there are no apparent landscape boundaries, the artificial influence of the boundary conditions on home range patterns can be mitigated by simulating the home ranges of adjacent packs in addition to the home range(s) of interest. As shown in figure 4.1, it is not necessary to have relocation data for peripheral packs. Simply specifying the locations of neighboring home range centers and using the model to simulate the movement and scent-marking behavior of individuals in these packs provides a method of predicting the position and shape of the home range boundaries between the packs of interest and adjoining packs.

The conspecific avoidance home range model formulation uses just two parameters to describe the movement and scent-marking behavior of individuals in all packs. As we shall see later (chapter 7), it is possible to formulate more detailed mechanistic home range models that include the effects of additional orientation cues such as landscape heterogeneity and resource availability on the movement behavior of individuals.

4.4. SUMMARY

In this chapter we formulated a mechanistic home range model in which individuals exhibit an avoidance response to foreign scent marks. Comparisons of the fit of this model to the fit obtained with the localizing tendency model developed in the previous chapter provide empirical support for the model's formulation of avoidance responses to foreign scent marks. Additionally, the results of the model fitting illustrate that having relocation data for individuals in adjacent groups is also important for accurately capturing the spatial arrangement of home range boundaries. We then showed how the conspecific avoidance model fits can be used to obtain predictions for individual movement and scent-marking behavior and to predict changes in home range patterns. The next two chapters explore the properties of this model in more detail.

CHAPTER FIVE

Comparative Analysis of Home Range Patterns Predicted by the Conspecific Avoidance Model

The results of the model–data fitting exercise in the previous chapter showed that a mechanistic home range model in which individuals exhibit an avoidance response to foreign scent marks was capable of capturing the macroscopic pattern of coyote home ranges at Hanford, Washington. In this chapter, we analyze in more detail the general properties and predictions of this conspecific avoidance model, using numerical simulations to explore its characteristics for the biologically realistic case of groups moving and interacting in two space dimensions. We then compare the predictions of the model to field observations of home range patterns and scent mark distributions in different carnivore species in different environments. This numerical analysis also serves as a prelude to the mathematical analysis in the subsequent chapter, which examines the model's properties for the idealized but analytically tractable case of two packs interacting in a single space dimension.

5.1. PREDICTED PATTERNS OF SPACE USE

The conspecific avoidance home range model reflects the macroscopic pattern of home ranges that results from a stochastic movement model in which (1) individuals exhibit an avoidance response to foreign scent marks, and (2) encounters with foreign scent marks cause individuals to scent-mark at a higher rate. In the previous chapter, we showed how the pattern of space use arising from the conspecific avoidance model for multiple groups interacting in two space dimensions is given by the following nondimensionalized system of equations:

Space use:

$$0 = \nabla^2 u^{(i)} - \nabla \cdot \left[\beta \vec{\mathbf{x}}_i u^{(i)} \sum_{j \neq i}^{n} p^{(j)} \right] \qquad (5.1)$$

Spatial distribution of scent marks:

$$0 = u^{(i)} \left[1 + m \sum_{j \neq i}^{n} p^{(j)} \right] - p^{(i)} \tag{5.2}$$

Boundary conditions:

$$\left[\nabla u^{(i)} - \beta u^{(i)} \vec{\mathbf{x}}_i \right] \cdot \vec{\mathbf{n}} = 0. \tag{5.3}$$

Examining the above equations, we see that the pattern of space use and the associated distribution of scent marks predicted by the conspecific avoidance model for any given spatial domain Ω are determined by the arrangement of home range centers $\mathbf{x}_i$ and the values of the two nondimensionalized parameters β and m. We now explore each of these in turn.

The Influence of Neighbors

A feature of the conspecific avoidance model critical for capturing the pattern of home ranges at Hanford is the dependency of a group's pattern of space use on the location of neighboring home ranges, which arises due to the coupling between equations (5.1) and (5.2). An example of this dependency is illustrated in color plate 6. When four groups are present in the region, the density of scent-marks is low and there is a high degree of overlap between home ranges (color plate 6a). Introduction of an additional group causes significant changes in the spatial pattern of scent marks and home ranges within the region (color plate 6b). The presence of the new home range causes a marked reorganization of the spatial arrangement of home ranges and scent marks; existing home range boundaries shift and new boundaries form between the new group and its neighbors (compare (a) and (b) in color plate 6). Overall, the reorganization that follows the addition leads to sharper and more heavily marked territorial boundaries, which reduces the level of home range overlap (see table 5.1).

The extent of home-range overlap between two territories can be quantified by the product of their probability density functions $u^{(i)}$ and $u^{(j)}$ integrated over the domain:

$$\text{overlap} = \int \int u^{(i)}(x, y) u^{(j)}(x, y) dx dy. \tag{5.4}$$

This is a weighted measure of overlap that accounts for the extent to which individuals within the two groups utilize different areas.

TABLE 5.1. Change in home range overlap between existing groups following the addition of a new group into the region shown in color figure 6. Numbers show the change in overlap as measured by the change in the value of equation (5.4) following the addition of the new group, with negative numbers indicating a decrease in overlap. The original extent of overlap between two groups is given in parentheses. The addition of the fifth group decreases overlap between all the existing groups.

	Group		
	1	2	3
1	—		
2	−0.140 (0.450)	—	
3	−0.058 (0.060)	−0.100 (0.220)	—
4	−0.039 (0.040)	−0.058 (0.060)	−0.140 (0.450)

The reorganization of home ranges seen in color plate 6 is similar to observed changes in home range patterns following the addition or removal of groups in natural populations. A nice example of this comes from Peterson's long-term study of wolves on Isle Royale, Michigan, where the formation of the Harvey Lake (HL) pack in 1980 led to a marked reorganization in the spatial arrangement home ranges on the island (figure 5.1).

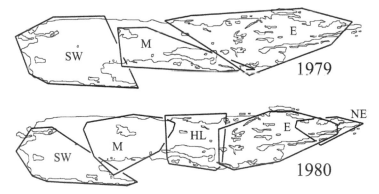

FIGURE 5.1. Wolf pack territories on Isle Royale, Michigan, 1979–80 (Peterson and Page 1988). Bold lines indicate the territory boundaries for the different packs. In 1979 there were three packs (SW, M and E) occupying territories on the island. The presence of the newly formed Harvey Lake pack (HL) in the central region of the island and the North Eastern pack (NE) at the eastern end of island caused shifts in the spatial arrangement of territories across the entire island.

Sensitivity of Movement to Foreign Scent

As we noted earlier, the β parameter of equation (5.1) is a measure of the strength of an individual's directed motion following encounters with foreign scent marks relative to its non-directed motion ($\beta = c/d = b\bar{\rho}^2/2\rho^2$; see chapter 4). Thus increases in the sensitivity of an individual's movement direction to the presence of foreign scent marks (larger b) or a larger ratio of mean to mean-square distance between turns (higher $\bar{\rho}^2/\rho^2$ ratio) increases β.

The effect of variation in β on the shape of the home ranges and the associated distribution of scent marks is shown in color plate 7. At low values of β, home ranges show a relatively high degree of overlap and the accompanying distribution of scent marks is relatively spread out, peaking in the interior of the home range (color plate 7a). Increasing the value of β causes space use to decrease more sharply in response to the presence of foreign scent marks. As a result, home ranges become less overlapping and the scent marks become concentrated at the edge of this well-defined territorial boundary (color plate 7b).

When individuals exhibit no directional movement ($\beta = 0$), equation (5.4) is unity, indicating complete overlap of the two home ranges. Positive values of β reduce overlap between the groups, the overlap decreasing as β increases. Overlap declines more rapidly when the scent-marking response m is high compared to when m is low (figure 5.2).

Thus when an individual's movements are increasingly dominated by a directed component of motion (increasing β), overlap between neighboring home ranges decreases (figure 5.2). The reduction in overlap causes the overall density of scent marks to decrease due to the reduced frequency of encounters with foreign scent marks, and also causes the scent marks to become increasingly concentrated at the boundary between home ranges (compare (a) and (b) in color plate 7).

This behavior of the model accords with the seasonal variation in home range patterns observed in many carnivore populations. Individuals localize around their dens following the birth of young (β high), causing home ranges to contract during the natal period, then increase their movements again as offspring become independent (β low), causing home ranges to expand following weaning (Harrison and Gilbert 1985; Person and Hirth 1991; Holzman and Conroy 1992). As in the model, associated with these changes in space use are changes in the distribution of scent marks (Bekoff and Wells 1982).

In addition, several studies have shown that seasonal declines in food availability cause increases in the daily movement rates of individuals. In the model, this corresponds to an increase in the mean-squared displacement of individuals that reduces the value of β. As color plate 7 shows, reductions in β cause home ranges to expand and become more overlapping, a prediction consistent with the findings of numerous studies, which have found that seasonal increases in

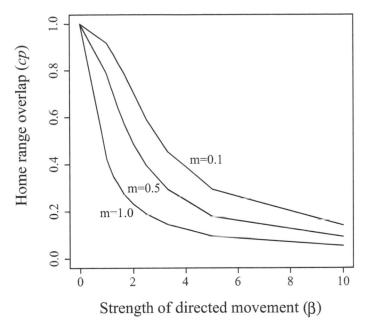

FIGURE 5.2. Effect of β parameter of the conspecific avoidance model on home range overlap. Plot shows how overlap changes as a function of the movement parameter β. When $\beta = 0$, equation (5.4) equals 1, indicating complete home range overlap. Increasing β reduces overlap between territories, overlap decreasing toward zero as β gets large. The decrease in home range overlap with increasing β is faster when the value of the overmarking parameter m is high and slows when m is low.

movement distances in response to declining food availability result in increased home-range overlap (Hoskinson and Mech 1976; Mech 1977; Packard and Mech 1980; Bekoff and Wells 1986; Peterson and Page 1988).

Strength of the Overmarking Response

The scent mark response parameter m reflects the extent to which individuals increase their scent-marking rate in response to encounters with foreign scent marks, relative to their background rate of marking (see equations (4.17) and (4.18) of the previous chapter). When an individual's marking rate is unaffected by foreign scent marks ($m = 0$), the pack's scent-mark distribution directly reflects its pattern of space use, with the highest densities of scent marks in the areas of highest space use around the home range center (color plate 8a). However, when individuals increase their marking rate in response to encounters with foreign scent marks ($m > 0$), the areas of highest scent-mark density are now found on the periphery of the home range (color plate 8b). This

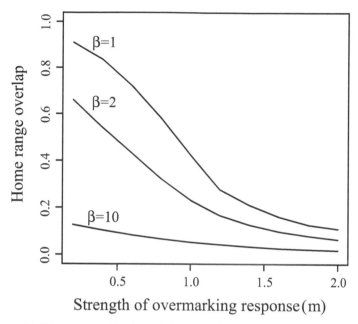

FIGURE 5.3. The scent-marking boundaries caused by increasing m reduces overlap between neighboring home ranges.

scent-mark boundary becomes increasingly well defined as m increases further (color plate 8c).

These changes in the spatial distribution of scent marks are accompanied by changes in the pattern of space use. Home-range overlap decreases with increasing m (Figure 5.3), the magnitude of the decline depending on the movement parameter β. At low values of β, when overlap between territories is high, increases in m cause marked declines in overlap between adjacent groups. When β is high, however, increases in m cause only relatively small decreases in home-range overlap (figure 5.3).

5.2. BORDER VERSUS HINTERLAND SCENT MARKING

The changing spatial patterns of scent marks that occur in response to changes in the overmarking response m (color plate 8) mirror an interesting continuum in the spatial distribution of scent marks in different carnivore populations. In some populations, including wolves in Minnesota (Peters and Mech 1975), coyotes in Wyoming and Alberta (Camenzind 1978; Bowen and McTaggart-Cowan 1980; Gese and Ruff 1996), lions (Schaller 1972), aardwolves (Richardson 1990), and spotted hyenas in Ngorongoro (Kruuk 1972), scent marks are concentrated around the periphery of the home range, forming an "olfactory bowl" (Figure 5.4a), while in other populations, including striped hyenas in the Serengeti (Kruuk 1976), brown and spotted hyenas in the Kalahari (Gorman and Mills 1984; Mills and Gorman 1987; Mills 1989), and coyote populations in Quebec (Barrette and Messier 1980), scent marks are concentrated in the home-range interior (Figure 5.4b).

Gorman and Mills (1984) termed these contrasting spatial distributions "border" and "hinterland" patterns of scent marking. The bowl-shaped distributions observed in wolves, coyotes, lions, and spotted hyenas in Ngorongoro are typical "border" marking patterns, while the interior distributions found in the brown hyenas and spotted hyenas in the Kalahari are characteristic "hinterland" marking patterns. Other populations, such as wolves in Riding Mountain National Park, Manitoba (Carbyn 1980; Paquet and Fuller 1990; Paquet 1991), exhibit intermediate patterns, suggesting that "border" and "hinterland" reflect two ends of a continuum of scent-mark spatial distributions found among carnivores. This idea is also supported by the distribution of scats within carnivore home ranges, which exhibit a similar range of spatial patterns (figure 5.5).

The simulations shown in color plate 8 suggest that the continuum of scent-mark spatial distributions seen in figures 5.4 and 5.5 may reflect differences in the scent-marking behavior of individuals, specifically, differences in the rate at which individuals scent-mark in response to encounters with foreign scent marks.

Differences in Behavior or Population Density?

While the variability in scent-mark distributions seen in figures 5.4 and 5.5 may be due to between-species differences in the overmarking response of individuals to foreign scent marks, an alternative explanation for the varying patterns of scent marks is that they are caused by differences in population density. The first hint of this is seen in color plate 6, where introduction of a new pack increased the level of scent marking on the boundaries between home ranges. However, the changes in space use and scent-mark patterns seen in this figure are rather complex, reflecting both a change in the spatial arrangement of home ranges and a change in the density of home ranges (number of packs occupying the region). We can separate out the effect of the change in density of home ranges by varying the area (A) of the domain over which we solve equations (5.1) and (5.2). From the nondimensionalization of section 4.2, we see that

$$\beta \propto \frac{1}{\sqrt{A}}, \quad \text{and} \quad m \propto \frac{1}{A}, \tag{5.5}$$

where $A = L^2$. Thus an increase in population density arising from an increase in density of home ranges causes simultaneous increases in β and m.

(a)

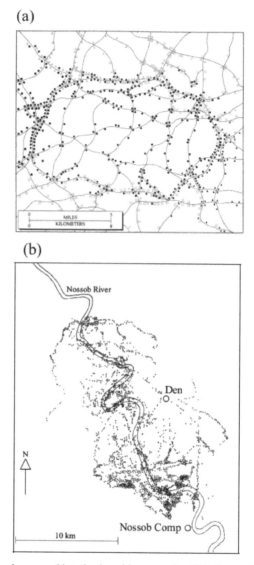

(b)

FIGURE 5.4. Border versus hinterland marking strategies. This figure shows the spatial distribution of scent marks within (a) a wolf (*Canis lupus*) territory in Minnesota (redrawn from Mech (1991)), and (b) a brown hyena *(Hyaena brunnea)* territory (redrawn from Gorman and Mills (1984)).

Alternatively, higher population densities can arise as a result of more individuals occupying each home range. Again from the nondimensionalization, we see that

$$\beta \propto N, \quad \text{and} \quad m \propto N, \tag{5.6}$$

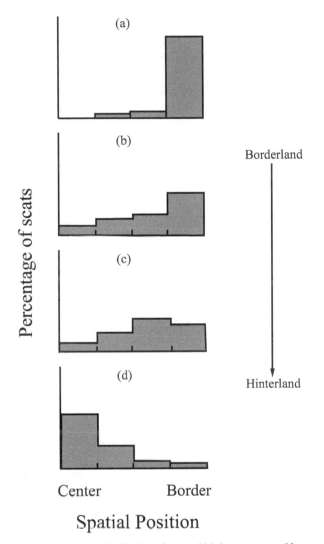

FIGURE 5.5. Observed spatial distribution of scats within home ranges of four carnivore species, (a) jackal, (b) badger (*Meles meles*), (c) fox (*Vulpes vulpes*), and (d) otter (*Lutra lutra*), reflecting the continuum of scent-mark patterns found within carnivore populations (redrawn from Macdonald 1980a). Variation in the strength of scent-marking response parameter m (eq. (5.2)) gives rise to similar pattern of variation (see color plate 8).

where N is the number of individuals within each group. Thus an increase in population density arising from an increase in group size causes a similar simultaneous increase in β and m. Note, however, that the scaling of the effect on β is different.

At low population density, β and m are low, resulting in overlapping home ranges and a hinterland distribution of scent marks that closely follows the pattern of space use (color plates 7a and 8a). At high population density, β and m increase, reducing home range overlap and causing the formation of a distinct scent-mark boundary (color plates 7b and 8c).

Thus changes in population density arising from either changes in the number of home ranges per unit area or changes in pack size, provide an alternative explanation for the borderland-hinterland continuum of marking patterns (figures 5.4 and 5.5). While differences in the overmarking response of individuals may account for the between-species differences seen in Figures 5.4 and 5.5, differences in population density provide a simple explanation for intraspecific differences in scent-mark patterns observed in some species. For example, as noted above, Gorman and Mills (1984) noted that spotted hyena home ranges in the Kalahari had hinterland distributions of scent marks, while in the Serengeti they had borderland distributions of scent marks. The authors attributed these differences in scent-mark patterns to differences in the movement and marking strategies of individuals in the different populations. However, as we have shown, the different spatial distributions of scent marks found by Gorman and Mills may simply reflect differences in either the density of home ranges or group size, rather than any change in the scent-marking behavior of the hyenas.

5.3. THE DISTRIBUTION OF SCENT MARKS ALONG BOUNDARIES

Another interesting feature in color plate 6 is the variability in the intensity of scent marking along home-range boundaries. Along boundaries where a neighboring home-range center is nearby, heavy scent marking occurs and home-range overlap is low, while along boundaries where the nearest home-range center is more distant, the density of scent marks is lower and overlap is higher. A similar pattern of variability is seen in the predicted scent-mark distributions at Hanford (see figure 4.3). This spatial variability in the intensity of scent marks around the edges of home ranges predicted by the conspecific avoidance model is consistent with observations by Kruuk (1978), who found that European badger *(Meles meles)* scent marks are more frequent on boundaries between nearby groups, where the potential for interaction between the groups is high and less frequent along boundaries with more distant groups, where the potential for interaction is lower (figure 5.6). Similar observations in a variety of other carnivores (Peters and Mech 1975; Kruuk 1978; Wells and Bekoff 1981; Bowen 1982) suggest that this association between the proximity of neighbors and the intensity of marking along different boundaries predicted by the conspecific avoidance model may be a general pattern that arises within carnivore populations.

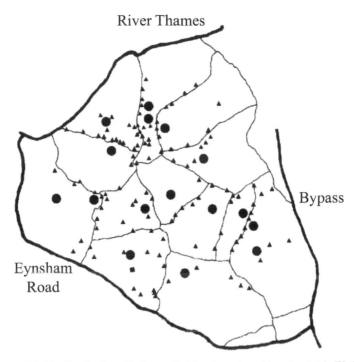

FIGURE 5.6. The distribution of badger setts (•) and scent-marking sites (▲) in Wytham Woods, Oxford. Consistent with the predictions of the conspecific avoidance model (color plate 6), boundaries between groups with nearby setts locations are more heavily marked than those with more distant neighbors (from Kruuk (1978)).

One versus Two Dimensions

Finally, before leaving the biologically realistic world of two dimensions and entering the world of one dimensional, analytically tractable solutions, we end with a cautionary reminder. Color plate 9 shows the patterns of space use and scent marks for four packs interacting on a symmetric, two-dimensional domain and the analogous solution for two packs interacting on a symmetric, one-dimensional domain. The spatial segregation of the packs in the one-dimensional solution is similar to that seen in the two-dimensional model solution. In both cases, the packs occupy distinct, relatively non-overlapping home ranges in which the movements of individuals are restricted by a boundary of foreign scent marks color plates 9a and b. The one-dimensional solution implies the formation of a scent-mark boundary between the two home ranges; however, it is only in the two-dimensional solution that we see that the scent marks are not broadly distributed around the edge of the home range but instead are concentrated into clusters along the axes between each pack's home-range center and those of its neighbors (color plate 9a, b). In other words, the clustering seen in the distribution of badger scent marks (figure 5.6) is a property of the conspecific avoidance model that only emerges when we solve the model in two dimensions.

5.4. SUMMARY

In this chapter we explored the predictions of the conspecific avoidance home range model for patterns of space use and the spatial pattern of scent marks under different levels of pack interaction, scent-induced avoidance response, and overmarking response by individuals. We began by simulating the effect of introducing a new home range into an existing landscape of contiguous home ranges. The simulation showed that adding an additional pack sharpens home-range boundaries between existing packs and between the new groups and their neighbors, thereby reducing home-range overlap. The changes seen in the simulation are qualitatively similar to the shifts in wolf home ranges on Isle Royale following the formation of the new pack on the island.

We then investigated the effect of varying the scent mark–induced movement bias parameter and the scent mark–induced marking response parameter. Changes in both parameters caused notable shifts in the shape of the home ranges, the spatial distribution of scent marks, and the level of home-range overlap. The changes in home-range overlap that accompany changes in the avoidance response of individuals resemble the seasonal changes in home-range overlap observed in many canid populations following changes in food availability. The differences in scent-mark distribution that accompany changes in the marking response of individuals to foreign scent marks parallel a continuum of scent-mark distributions found in different carnivore populations. However, further analysis shows that these different scent-mark patterns could also be due to differences in population density. Finally, an analysis of the spatial distribution of scent marks within a badger population shows that the characteristic clustering of scent-mark "hot spots" on the periphery of badger territories can be understood in terms of distance to the closest competing group, a feature of the model that only becomes apparent when we examine its properties in two space dimensions.

CHAPTER SIX

Mathematical Analysis
of the Conspecific Avoidance Model

In this chapter, we examine the conspecific avoidance model implemented for an idealized case of two packs moving and interacting in one space dimension. While this means an inevitable loss of biological realism, the pairwise one-dimensional case retains many of the qualitative features of the two-dimensional, multiple-group case investigated in the two previous chapters. Following the approach of Lewis et al. (1997), we use the analytical tractability of the idealized model to gain mathematical insight into the properties and predictions highlighted by the numerical simulations in the previous chapter (see also White et al. (1998) and Murray (2002)). We also use this idealized case to examine the consequences of some alternative functional forms for the scent-marking and avoidance behavior of individuals upon patterns of space use and scent marking. As in chapter 2, we invite non-mathematical readers to skip to section 6.5, where we summarize the main findings of our analysis.

6.1. MODEL EQUATIONS

Solutions of the nondimensionalized conspecific avoidance model equations (4.15)–(4.18) for a symmetric, pairwise interaction between two groups whose home-range centers are located at the left and right ends of a one-dimensional domain (x) of unit length, are given by the following nondimensionalized, time-independent equations:

$$0 = d\frac{\partial^2 u}{\partial x^2} - \frac{\partial}{\partial x}\left[qu\right], \tag{6.1}$$

$$0 = d\frac{\partial^2 v}{\partial x^2} + \frac{\partial}{\partial x}\left[pv\right], \tag{6.2}$$

$$0 = u(1 + mq) - p, \tag{6.3}$$

$$0 = v(1 + mp) - q, \tag{6.4}$$

with boundary conditions

$$d\frac{\partial u}{\partial x} - cqu = 0, \qquad d\frac{\partial v}{\partial x} + cpv = 0, \qquad \text{at} \quad x = 0, 1 \qquad (6.5)$$

and conservation conditions

$$\int_0^1 u(x)\, dx = \int_0^1 v(x)\, dx = 1. \qquad (6.6)$$

6.2. IMPACT OF THE SCENT-MARKING RESPONSE

The numerical investigations in chapter 5 suggest that observed differences in patterns of space use and scent marking within carnivore populations can be explained by differences in the magnitude of the scent-marking parameter m arising from either (1) changes in the magnitude of an individual's scent-marking response to encounters with foreign scent marks or (2) changes in the density of home ranges (see figures 5.3–5.5 and color plate 7). In this chapter we use an idealized implementation of the conspecific avoidance model to investigate this result mathematically, considering the relationship between the m parameter and patterns of space use and the spatial distribution of scent marks.

Case 1: No Marking Response to Foreign Scent Marks

We first consider the case in which there is no change in an individual's marking response to encounters with foreign scent marks ($m = 0$, see figure 6.1a). Equations (6.3) and (6.4) imply that

$$p = u \qquad \text{and} \qquad q = v, \qquad (6.7)$$

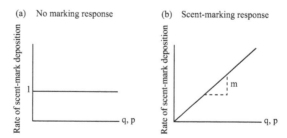

FIGURE 6.1. Functional forms for the scent-marking response of individuals in relation to the density of foreign scent marks (p or q depending on the pack). (a) No response ($m = 0$ in eqs. (6.3) and (6.4)). (b) Linear response ($m > 0$ in eqs. (6.3) and (6.4)).

i.e., the scent-mark density of the two packs directly reflects their patterns of space use. Substituting equation (6.7) into equations (6.1) and (6.2), integrating (6.1)–(6.2) with respect to space x, then applying the zero-flux boundary conditions equation (6.5), implies

$$\frac{\partial u}{\partial x} = -\frac{c}{d}uv, \qquad \frac{\partial v}{\partial x} = \frac{c}{d}uv. \tag{6.8}$$

Adding these equations implies that

$$\frac{\partial}{\partial x}(u+v) = 0. \tag{6.9}$$

Applying the integral constraint equation (6.6), this implies

$$u(x) + v(x) = 2 \tag{6.10}$$

at all points on the domain $0 \leq x \leq 1$. The solutions of equation (6.8), given equation (6.10), are two logistic equations with space as the independent variable:

$$u_x = -\frac{c}{d}u\,(2-u), \quad v_x = \frac{c}{d}v\,(2-v). \tag{6.11}$$

Thus in the absence of an overmarking response to foreign scent marks, space use declines logistically from the home-range centers of the two packs (figure 6.2a).

Case 2: Marking Response to Foreign Scent Marks

When individuals engage in overmarking ($m > 0$), equations (6.3) and (6.4) yield

$$p = \frac{u(1+mv)}{1-m^2uv}, \tag{6.12}$$

$$q = \frac{v(1+mu)}{1-m^2uv}, \tag{6.13}$$

and substitution into equations (6.1) and (6.2) yields

$$0 = c\frac{uv(1+mu)}{1-m^2uv} + du_x, \tag{6.14}$$

$$0 = -c\frac{uv(1+mv)}{1-m^2uv} + dv_x. \tag{6.15}$$

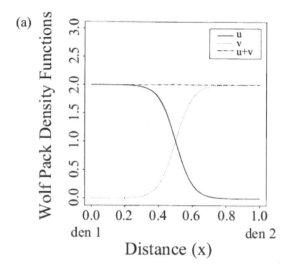

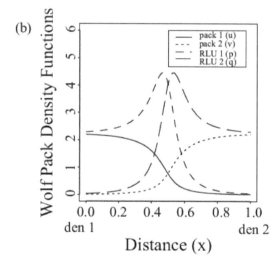

FIGURE 6.2. Home range patterns arising from the conspecific avoidance model (eqs. (6.1)–(6.4)), implemented for a pairwise interaction between two groups in a single space dimension x with home ranges located at opposing ends of a $[0, 1]$ domain. (a) Case 1: In the absence of a scent-marking response to encounters with foreign scent marks ($m = 0$, figure 6.1a), space use by the two groups, indicated by the curves $u(x)$ and $v(x)$, is given by solution of a pair of logistic equations (6.11). Note that the level of scent marking by the groups, indicated by the curves $p(x)$ and $q(x)$, directly reflects their respective intensities of space use given by the curves $u(x)$ and $v(x)$ (see eq. (6.7)). Note also that $u(x) + v(x) = 2$ point-wise across the domain, implying that the total intensity of space use is spatially uniform (see eq. (6.10)). (b) Case 2: When individuals increase their scent-marking rate in response to encounters with foreign scent marks ($m > 0$, figure 6.1b), space use by the two groups, indicated by the curves $u(x)$ and $v(x)$, is given by the solution of equation (6.19). Note the "bowl-shaped" scent-mark densities $p(x)$ and $q(x)$ for the two groups, which are the one-dimensional analog of the patterns shown in color plate 8.

Defining

$$\Gamma(w) = \int_0^w \frac{d\omega}{1 + m\omega} = \log(1 + mw)/m,$$

we observe from equations (6.14)–(6.15) that

$$\Gamma(u) + \Gamma(v) = \Gamma(u(0)) + \Gamma(v(0)) = k \quad \text{(a constant)}. \tag{6.16}$$

Thus

$$(1 + mu)(1 + mv) = \exp(mk) \tag{6.17}$$

describes u in terms of v and vice versa.

Substitution of equation (6.17) into equation (6.15) yields

$$0 = -c \frac{v \left(\frac{\exp(mk)}{1+mv} - 1 \right)/m}{1 - mv \left(\frac{\exp(mk)}{1+mv} - 1 \right)} + \frac{dv_x}{1 + mv}, \tag{6.18}$$

while substitution into equation (6.14) yields a similar equation for u. Equation (6.18) is separable, and integration yields an implicit formula for $v(x)$:

$$\frac{v(x)}{(E - mv(x))(1 + mv(x))^E} = \frac{v(0) \exp\{\frac{cEx}{dm}\}}{(E - mv(0))(1 + mv(0))^E}, \tag{6.19}$$

where $E = \exp(mk) - 1$. The similar equation for u can be obtained by the same method. For some cases the solution can be calculated explicitly. For example, if $E = 1$, then

$$v(x) = \frac{1}{2m^2 v(0)} \left(- \exp \left\{ \frac{-cx}{dm} \right\} (1 - m^2 v^2(0)) \right.$$

$$\left. + \sqrt{\left(\exp \left\{ \frac{-cx}{dm} \right\} (1 - m^2 v^2(0)) \right)^2 + 4m^2 v^2(0)} \right). \tag{6.20}$$

Finally, the initial conditions $u(0)$ and $v(0)$ must be chosen so as to satisfy the integral constraints (eq. (6.6)).

We use equation (6.17) to observe that

$$0 = \frac{v_x}{1 + mv} + \frac{u_x}{1 + mu}. \tag{6.21}$$

This can be used to simplify expressions for p_x derived from equation (6.12) and q_x derived from equation (6.13):

$$p_x = \frac{(1+mu)(mu-1)}{(1-muv)^2} v_x$$

$$q_x = \frac{(1+mv)(mv-1)}{(1-muv)^2} u_x.$$

Because the home range centers are located at opposing edges of the domain, the equations for space use $u(x)$ and $v(x)$ have no critical points (see color plate 9b). Proof of this assertion can be found in Lewis et al. (1997). Given this, possible interior maximums for $p(x)$ and $q(x)$ are when $u(x) = 1/m$ and $v(x) = 1/m$, respectively. Hence there is an interior maximum for p if and only if $u(0) \geq 1/m \geq u(1)$, and there is an interior maximum for $q(x)$ if and only if $v(0) \leq 1/m \leq v(1)$. In other words, if the behavioral response function m is sufficiently steep, then $1/m$ will lie in the above interval and scent-mark peaks will arise on the boundary of the home ranges. Solutions of the model equations confirm this result (see figure 6.2b).

6.3. EXISTENCE OF A BUFFER ZONE

Figure 6.3 shows total space use $(u(x) + v(x))$ and the total scent-mark density $(p(x) + q(x))$ across the region. In contrast to the $m = 0$ case, where these were spatially uniform (figure 6.2b), when $m > 0$ total space use drops near the territorial boundary located at $x = 1/2$, and the total scent-mark density is highest (figure 6.3). This property of the model accords with empirical observations of wolves in northeastern Minnesota, where contiguous home ranges are separated by areas of reduced space use, known as "buffer zones" (Mech 1977; Mech 1994).

We now derive the conditions under which the solutions of the scent-marking model (eqs. (6.1)–(6.4)) result in a "buffer zone" between the two groups. Mathematically, a buffer zone corresponds to the existence of an interior minimum in the sum of the two probability density functions $u(x) + (v)$. From equations (6.1)–(6.4) we observe that

$$\frac{\partial u}{\partial x} = -\frac{1}{d}qu, \tag{6.22}$$

$$\frac{\partial v}{\partial x} = \frac{1}{d}pv, \tag{6.23}$$

$$p = u(1 + mq), \tag{6.24}$$

$$q = v(1 + mp), \tag{6.25}$$

subject to the integral constraints (eq. (6.6)).

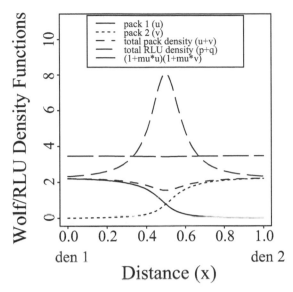

FIGURE 6.3. The cumulative expected space use $(u(x) + v(x))$ and cumulative scent mark density $(p(x) + q(x))$ for Case 2 (figure 6.2b). The function $(1 + mu)(1 + mv)$ is constant across the domain, as predicted by equation (6.17). Note the elevated scent-mark density and buffer zone at $x = 0.5$.

Since we are considering an interaction between two identical groups with home range centers at opposing ends of the domain, the solution to equations (6.22)–(6.25) is invariant when $x \to 1 - x$ and $u \leftrightarrow v$ and $p \leftrightarrow q$. Thus the solution of the equations is symmetric about the midpoint $x = 0.5$, i.e., at $x = 0.5$,

$$u = v, \quad p = q,$$

$$\frac{\partial u}{\partial x} = -\frac{\partial v}{\partial x}, \quad \frac{\partial p}{\partial x} = -\frac{\partial q}{\partial x}, > 0.$$

Thus

$$\frac{\partial}{\partial x}(u + v) = 0,$$

$$\frac{\partial q}{\partial x} = \frac{\frac{\partial v}{\partial x}(1 + m(p))}{1 + vm'(p)} > 0,$$

and

$$(u + v)_{xx} = \frac{1}{d}\{pv - qu\}_x,$$

$$= \frac{2}{d}\{up_x - pu_x\},$$

$$= \frac{2u^2}{d}\frac{\partial}{\partial x}\left\{\frac{p}{u}\right\} = \frac{2u^2}{d}\frac{\partial}{\partial x}\{(1 + mq)\},$$

$$= \frac{2u^2}{d}mq_x, \tag{6.26}$$

which is always positive for $m > 0$, indicating that total space use $(u + v)$ reaches a minimum at $x = 0.5$, implying a buffer zone midway between the two interacting groups.

6.4. GENERALIZED RESPONSE FUNCTIONS

When formulating the conspecific avoidance model, we assumed for simplicity that both the avoidance response and the scent-marking response of individuals were linear functions of foreign scent-mark density. While these responses may be approximately linear over a certain range of foreign scent-mark densities, the assumption of linear responses may not be biologically realistic over the range of scent-mark densities encountered, since it implies that in areas of high scent-mark density, arbitrarily high rates of directed movement and scent marking are possible. Below we investigate the consequences of relaxing this assumption by analyzing alternative formulations in which the movement and scent-marking responses of individuals to foreign scent marks are bounded.

Case 3: Bounded Avoidance Response

We consider a simple case of a bounded avoidance response in which individuals exhibit a step response in their avoidance behavior. Suppose when the density of foreign scent marks is below a critical value $(q_c = p_c)$ individuals have no avoidance response, but above this level individuals exhibit an avoidance response of fixed magnitude c_{max}. Under this assumption, the equations for space use (6.1)–(6.2) are replaced by

$$0 = d\frac{\partial^2}{\partial x^2} - \frac{\partial}{\partial x}\left[c_{max}H(q - q_c)u\right], \tag{6.27}$$

$$0 = d\frac{\partial^2}{\partial x^2} + \frac{\partial}{\partial x}\left[c_{max}H(p - p_c)v\right], \tag{6.28}$$

where $H(\cdot)$ is the Heaviside (step) function (figure 6.4a). The scent-mark equations (6.3) and (6.4) are unchanged.

Using symmetry, we can construct a piecewise solution to equations (6.27)–(6.28). Defining x_c as the position at which the scent-mark density $q = q_c$, then

$$u = \begin{cases} u(0) & \text{if } 0 \leq x \leq x_c, \\ u(0)\exp\left(-\frac{c_{max}}{d}(x - x_c)\right) & \text{if } x_c < x \leq 1. \end{cases} \tag{6.29}$$

$$v = \begin{cases} u(0) & \text{if } 1 - x_c \leq x \leq 1, \\ u(0)\exp\left(-\frac{c_{max}}{d}(1 - x - x_c)\right) & \text{if } 0 \leq x < 1 - x_c. \end{cases} \tag{6.30}$$

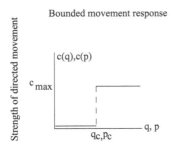

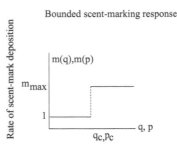

FIGURE 6.4. Bounded movement and scent-mark response functions. (a) Case 3: Individuals exhibit a step response in their avoidance movement behavior in relation to increasing foreign scent-mark density (eqs. (6.27) and (6.28)). (b) Case 4: Individuals exhibit a step response in their scent-marking behavior in relation to foreign scent mark density (eqs. (6.33) and (6.34)).

We can then use the integral constraint (eq. 6.6), to obtain an expression for $u(0)$ in terms of x_c:

$$u(0)\left(x_c + \frac{d}{c_{max}}\left(1 - \exp\left(\frac{c_{max}}{d}(x_c - 1)\right)\right)\right) = 1. \qquad (6.31)$$

The values of p and q are given by equations (6.12) and (6.13). Using equation (6.13) and equations (6.29)–(6.30), we calculate

$$q_c = q(x_c) = \frac{v(x_c)(1 + \mu u(0))}{1 - m^2 u(0) v(x_c)}, \qquad (6.32)$$

$$v(x_c) = u(0) \exp\left(-\frac{c_{max}}{d}(1 - 2x_c)\right).$$

Simultaneous solution of equations (6.31) and (6.32) yields the maximum expected density $u(x)$ in terms of the parameters d, c_{max}, q_c, and m.

The patterns of space use and scent-mark distribution that arise from individuals exhibiting a step response in their avoidance behavior are shown in figure 6.5. While a buffer zone is evident in the original linear formulation (figure 6.3b), the step-avoidance response greatly enhances the size of the buffer zone (figure 6.5). The quantity $(1 - 2x_c)$ gives a measure of the width of the buffer zone. For any given width, equation (6.32) implies that the depth of the buffer zone is an increasing function of the ratio of directed to random motion in the presence of foreign scent marks c_{max}/d.

Case 4: Bounded Scent-Marking Response

In a similar manner, we now examine the consequences of a bounded scent-marking response, in which individuals exhibit a step response in their scent-marking behavior. Suppose individuals show no change in their marking

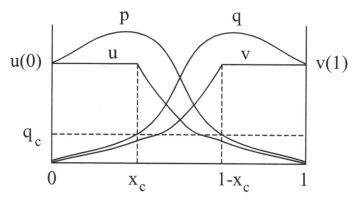

FIGURE 6.5. Analytical solution to the scent mark avoidance model with a bounded movement response (eqs. (6.22)–(6.28)). The patterns of space use for the two groups $u(x)$ and $v(x)$ are given by eqs. (6.29)–(6.30), with $u(0)$ and x_c determined by eqs. (6.31)–(6.32). The corresponding spatial distribution scent marks $p(x)$ and $q(x)$ are given by eqs. (6.12)–(6.13). Note the wide buffer zone between the two groups.

rate following encounters with foreign scent marks that are below a threshold density $q_m = p_m$, and that above this threshold individuals mark at rate m_{max} (figure 6.4b). Assuming the original form for avoidance response of individuals, the equations for space use (6.1)–(6.2) are unchanged; however, the scent-mark equations (6.3)–(6.4) are replaced by

$$0 = u(1 + m_{max}H(q - q_m) - p, \qquad (6.33)$$

$$0 = v(1 + m_{max}H(p - p_m) - q, \qquad (6.34)$$

where $H(\cdot)$ is the Heaviside function (figure 6.4b).

Figure 6.6 shows the solution to equations (6.1)–(6.2) and (6.33)–(6.34) in terms of u and v. Note that if (u, v) lies within the shaded region of figure 6.6 there is not a unique time-independent solution. However, outside of this region in $u-v$ phase space, the solution can be calculated in each sub region in a manner similar to that used in Case 3 above. At the transition from region B to region D the expected scent-mark density for pack 1 $(p(x_m))$ jumps from $u(x_m)$ to $(1 + m_{max})u(x_m)$, and at the transition from region D to region C the expected scent-mark density for pack 2 $(q(1 - x_m))$ jumps from $(1 + m_{max})v(1 - x_m)$ to $v(1 - x_m)$ (figure 6.6). Using symmetry, the relationship between $u(x)$ and $v(x)$ given in terms of $k = u(0) + (1 + m_{max})v(0) - m_{max}q_m$ is

$$
\begin{aligned}
u + v + m_{max}v &= k + m_{max}q_m && \text{if} && 0 \le x \le x_m, \\
u + v &= k && \text{if} && x_m < x < 1 - x_m, \\
u + v + m_{max}u &= k + m_{max}q_m && \text{if} && 1 - x_m \le x \le 1.
\end{aligned}
\qquad (6.35)
$$

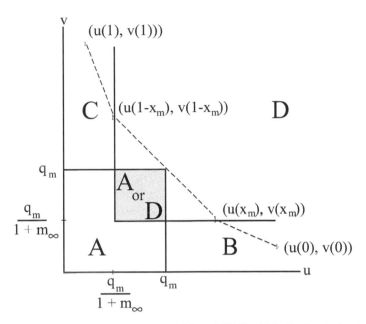

FIGURE 6.6. Solution to equations (6.1)–(6.2) and (6.33)–(6.34). Region A: $(p, q) = (u, v)$, region B: $(p, q) = (u, (1 + m_{max})v)$, region C: $(p, q) = ((1 + m_{max})u, v)$, and region D: $(p, q) = ((1+m_{max})u, (1+m_{max})v)$. Note that if $q_m/(1+m_{max}) < u, v < q_m$, then the equilibrium cannot be determined uniquely, being either case A or D above. If parameters are chosen so that $k = u(0) + (1+m_{max})v(0) - m_{max}q_m > q_m$ (eq. (6.35)), then the indeterminate region is bypassed in phase space. A sample solution is shown, with the three parts of the solution given by equation (6.35).

The value of k is chosen so as to satisfy the integral constraints (eq. (6.6)). A sample solution is shown in figure 6.6 and the corresponding spatial pattern of space use and scent-marking is shown in figure 6.7. Here, the step function in the scent-marking response of individuals yields a solution which is discontinuous in p and q. In other words, a direct consequence of the scent-marking step response is an abrupt jump in p and q at the edge of the buffer zone, implying a sharp "lip" on the bowl-shaped distribution of scent marks (figure 6.7).

6.5. SUMMARY

Mathematical analysis in a single space dimension gives analytic insight into the qualitative patterns of space use and scent marking predicted by the conspecific avoidance home-range model. When individuals exhibit an avoidance response to foreign scent marks but no overmarking response to encounters with foreign scent marks, this results in logistic patterns of space use for each pack,

78   MATHEMATICAL ANALYSIS

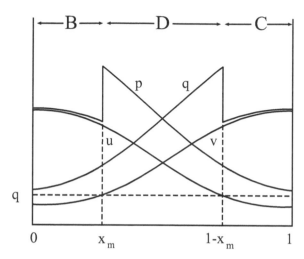

FIGURE 6.7. Analytical solution of the conspecific avoidance home range model (eqs. (6.1)-(6.1) and (6.33)–(6.34)) when individuals exhibit a switching response in their scent-mark behavior in response to foreign scent mark density (figure 6.4b). The corresponding solution trajectory is given in figure 6.6. The solution method is the same as described in section 6.2., except with the relationship between $u(x)$ and $v(x)$ in each subregion given by equation (6.35). Note the abrupt jumps in $p(x)$ and $q(x)$.

and no region of reduced total space use (buffer zones) between neighboring home ranges. In addition, the spatial distribution of scent marks for each group directly reflects their pattern of space use. The inclusion of an over marking response to foreign scent marks results in the formation of buffer zones between neighboring groups and, if the magnitude of the overmarking response is sufficiently large, also gives rise to a bowl-shaped distribution of scent marks within the home range. The conspecific avoidance model thus provides a simple mechanistic explanation for the buffer zones and patterns of scent marks observed in Minnesota wolf populations. When individuals exhibit a step, rather than linear, avoidance response to encounters with foreign scent marks, this can give rise to wide buffer zones between groups. Finally, when individuals exhibit a step overmarking response to foreign scent marks, this can give rise to a distinct "lip" in the distribution of scent marks on the edge of the home range.

CHAPTER SEVEN

The Influence of Landscape
and Resource Heterogeneity on Patterns
of Space Use

While the conspecific avoidance model analyzed in the preceding chapters captures the influence of neighbors on home-range patterns within a region, the landscape in which the animals move was assumed to be entirely homogeneous. For the relatively uniform sagebrush environment at Hanford this may not be an unreasonable assumption (see color plate 10a). However, many environments such as the Lamar Valley region in Yellowstone National Park (color plate 10b) are heterogeneous mosaics of different terrain and habitat types whose characteristics significantly alter the movement behavior of individuals. These changes in movement behavior can arise either as a direct response to changes in the physical landscape such as terrain steepness or snow depth, or as an indirect response to the changing biological properties of the landscape, such as a shift in foraging behavior in response to changes in prey availability in different habitats. In this chapter, following the analysis of Moorcroft et al. 2006, we illustrate how these kinds of spatial heterogeneities can be incorporated into mechanistic home range models by developing two extensions to the conspecific avoidance model analyzed in chapters 4–6 that account for the influence of landscape heterogeneity and foraging responses to changing resource availability.

7.1. LANDSCAPE HETEROGENEITY

Color plate 11 shows the spatial distribution of coyote home ranges in the Lamar region of Yellowstone National Park prior to the reintroduction of wolves in 1995. As at Hanford, the coyotes in Lamar during this period occupied distinct, relatively non-overlapping home ranges. However, unlike the Hanford landscape, which is relatively flat, the Lamar region consists of a valley surrounded by mountains up to 850 meters higher than the valley floor (see color plate 10). As can be seen from the spatial distribution of relocations in color plate 11, there is a clear influence of this topography on patterns of space use in the region, with coyotes confining their movements almost exclusively to lower elevation areas. Not surprisingly, because of this heterogeneous terrain, the conspecific avoidance home-range model that was able to characterize the Hanford dataset (eqs. (4.20)–(4.21)) gives a poor fit to the Lamar Valley relocation data (color plate 12).

We can account for this influence of topography using a similar approach to the one we employed in chapter 4 to incorporate the influence of foreign scent marks on patterns of movement, that is, by specifying how spatial variability in terrain steepness alters an individual's redistribution kernel describing its fine-scale movement behavior. Specifically, we consider how terrain steepness alters an individual's probability of moving in a given direction.

In our original formulation of the conspecific avoidance (CA) home-range model, individuals had a non-uniform circular distribution of movement directions that reflected an avoidance response to foreign scent marks. There are a variety of assumptions we might make regarding the combined influence of scent marks and terrain on the movement behavior of individuals. For simplicity, we consider the case where their effects are additive. An individual's distribution of movement directions is now a weighted sum of two separate non-uniform circular distributions: $K_\tau = \psi K_\tau^P + (1 - \psi)K_\tau^S$ where the distributions K_τ^P and K_τ^S respectively represent the influence of foreign scent marks and terrain steepness on an individual's probability of moving in a particular direction.

Following the approach of section 4.1, the movement kernel for an individual from pack i is

$$k(\mathbf{x'}, \mathbf{x}, \tau, t) = \frac{1}{\rho} f_\tau(\rho) \left(\psi K_\tau^P(\phi, \widehat{\phi}_H) + (1 - \psi)K_\tau^S(\phi, \widehat{\phi}_{z,}) \right). \qquad (7.1)$$

Here the kernel K_τ^P describes bias toward the den, situated at $\mathbf{x}_H$, based on level of total foreign marks encountered at a given location $P(\mathbf{x}, t) = \sum_{j \neq i}^{n_{pack}} p^{(j)}(\mathbf{x}, t)$. The kernel K_τ^S describes bias toward lower elevation areas based on the steepness of the terrain $S(\mathbf{x})$, whose value is given by the magnitude of the elevation gradient: $S(\mathbf{x}) = |\nabla z(\mathbf{x})|$. The quantity $\widehat{\phi}_H$ (identical to $\widehat{\phi}$ in section 4.1) is given as the angle between $\mathbf{x}$ and the den at $\mathbf{x}_H$: $\tan^{-1}((y - y_H)/(x - x_H))$. The quantity $\widehat{\phi}_z$ is given as the angle of the downhill slope at point $\mathbf{x}$: $\tan^{-1}((\partial z/\partial y)/(\partial z/\partial x))$. The kernels K_τ^P and K_τ^S are taken to be von Mises distributions (eq. (3.2)), as in section 4.1, but with concentration parameters

$$\kappa_\tau^P = b\bar{\rho}_\tau P(\mathbf{x}, t), \qquad (7.2)$$

$$\kappa_\tau^S = a\bar{\rho}_\tau S(\mathbf{x}). \qquad (7.3)$$

Substituting these into equation (2.16), methods similar to those in appendix E can be used to simplify the resulting equations. The probability density function $u(\mathbf{x}, t)$ for the expected space use of pack i is given by the partial differential equation

$$\frac{\partial u^{(i)}}{\partial t} = d\nabla^2 u^{(i)} - \nabla \cdot \left[c_p u^{(i)} \vec{\mathbf{x}}_i P \right] + \nabla \cdot \left[c_z u^{(i)} \nabla z \right] \qquad (7.4)$$

where $\vec{\mathbf{x}}_i$ is a unit vector pointing toward the den site for pack $U^{(i)}$. The scent-mark advection coefficient c_p, terrain advection coefficient c_z and diffusion coefficient d are given by

$$c_p = \lim_{\tau \to 0} \frac{\psi b \bar{\rho}_\tau^2}{2\tau}, \quad c_z = \lim_{\tau \to 0} \frac{(1 - \psi)a\bar{\rho}_\tau^2}{2\tau}, \quad \text{and} \quad d = \lim_{\tau \to 0} \frac{\bar{\rho}_\tau^2}{4\tau}. \qquad (7.5)$$

Note that the case $\Psi = 1$ corresponds to the CA model.

Equation (7.4) can be non-dimensionalized, yielding the following equations for the expected steady-state pattern of space use:

$$0 = \underbrace{\nabla^2 u^{(i)}}_{\text{random motion}} \quad \underbrace{\nabla \cdot \left[\beta \vec{\mathbf{x}}_i u^{(i)} \sum_{j \neq i}^{n} p^{(j)} \right]}_{\text{scent-mark avoidance}}$$

$$+ \quad \underbrace{\nabla \cdot \left[\alpha_z u^{(i)} \nabla z \right]}_{\text{avoidance of steep terrain}} \quad , \quad i = 1 \ldots n, \qquad (7.6)$$

where β is the strength of the scent-mark avoidance relative to the random component of motion, and α_z is the strength of the steep terrain avoidance relative to the random component of motion (compare equation (7.6) with equation (4.20)). The equations describing the spatial distribution of scent marks (eqs. (4.21)) are unchanged:

$$0 = u^{(i)} \left[1 + m \sum_{j \neq i}^{n} p^{(j)} \right] - p^{(i)}, \quad i = 1 \ldots n. \qquad (7.7)$$

We label equations (7.6)–(7.8) the "steep-terrain avoidance plus conspecific avoidance" (STA+CA) model. Like the CA model, equations (7.6)–(7.8) have associated zero-flux boundary conditions:

$$\left[\nabla u^{(i)} - \beta u^{(i)} \vec{\mathbf{x}}_i + \alpha_z u^{(i)} \nabla z \right] \cdot \vec{\mathbf{n}} = 0, \quad \text{on} \quad \partial\Omega, \qquad (7.8)$$

and are subject to the following integral constraint:

$$\int_\Omega u^{(i)} dx = 1. \qquad (7.9)$$

TABLE 7.1. Details of the home range model fits (figures 7.3 and 4) to relocation data for five contiguous packs in Lamar Valley, Yellowstone National Park ($n_{total} = 1955$ relocations). Parameter values, log-likelihood scores ($l(\theta)$), and Akaike Information Criterion (AIC) scores are given for the fit of the conspecific avoidance (CA) home range model (eqs. (4.20)–(4.21)), the fit of the steep terrain avoidance plus conspecific avoidance (STA+CA) home range model (eqs. (7.14) and (7.7)), and the fit of the prey availability plus conspecific avoidance (PA+CA) model (eqs. (7.14) and (7.7)). n_{total} is the total number of data points used in each model fit $n_{total} = \sum_{i=1}^{n} n_{reloc}^{(i)}$ (see eq. (4.23)). In all three models, β indicates the ratio of directed movement per unit of scent-mark density encountered relative to the strength of non-directed movement (see eqs. (4.19), (7.11), and (7.14)) and m governs the sensitivity of an individual's marking rate to foreign scent marks (eqs. (4.21) and (7.7)). The parameter α_z of the STA+CA model governs the magnitude of steep terrain avoidance relative to the strength of non-directed movement (see eq. (7.6)). In the PA+CA model, α_r governs the sensitivity of turning frequency to resource density (see eq. (7.14)).

	CA	STA+CA	PA+CA
Equations	4.20–4.21	7.6 & 7.7	7.14 & 7.7
Parameters			
β	2.86	3.31	3.35
m	0.047	0.34	0.0030
α_z	–	30.8	–
α_r	–	–	0.265
$l(\theta)$	2453.5	3296.2	3753.1
AIC	−4903.0	−6586.4	−7500.2

Color plate 13 shows the fit of the STA+CA model to the Lamar relocation dataset collected by Crabtree et al. The inclusion of a steep-terrain avoidance response greatly improves the ability of the home range model to characterize home range patterns in the Lamar region, resulting in home ranges that, like the observations, are confined to the lower part of the valley (color plate 13). As a result, the STA+CA model gives a substantially improved goodness-of-fit over the CA model fit shown in color plate 12 ($\Delta AIC = 1683.4$ and $\Delta l = 842.7$, $1 d.f.$, $p < 0.0001$; see table 7.1).

7.2. RESOURCE HETEROGENEITY AND FORAGING BEHAVIOR

Another widespread cause of spatial variability in movement behavior of carnivores arises from the impact of heterogeneity in resource availability on foraging behavior. Indeed, the effect of topography on coyote space use in Lamar Valley seen in color plate 13 is, in large part, likely to be a result of coyotes preferring to forage in low-elevation habitats, where resource densities

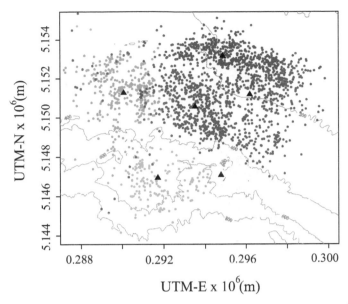

COLOR PLATE 1 (see chapter 3) Radio-tracking data for six contiguous coyote packs at Hanford ALE collected by Crabtree (1989). Colored points (•) indicate radio locations for individuals belonging to different packs. ▲ indicates centroid of the relocation coordinates for each pack, used in the model as the home range center.

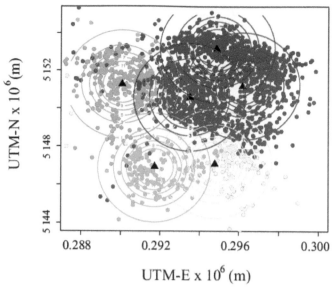

COLOR PLATE 2 (see chapter 3) Contour lines showing regional fit of equation (3.7) to relocation data (•) for six coyote packs at Hanford ALE collected by Crabtree (1989) (2325 relocations). As in figure 3.9, the contour interval for $u(x, y)$ is 2 and the home range centers for each pack are shown (▲). Maximum likelihood estimate for β is 17.0.

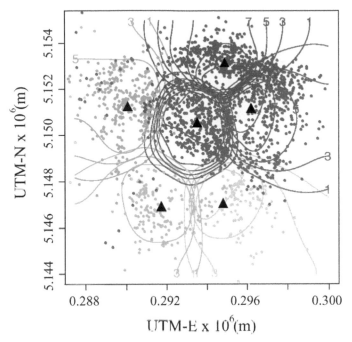

COLOR PLATE 3 (see chapter 4) Regional of fit the conspecific avoidance model showing home ranges for all six coyote groups in the study region obtained by fitting equations (4.20)–(4.21) to relocation data (•) for all six coyote groups at Hanford ALE collected by Crabtree (1989). The contour interval for $u(x, y)$ is 2, and the home range centers for each pack are shown (▲). Maximum likelihood values and estimates for β and m are given in table 4.1.

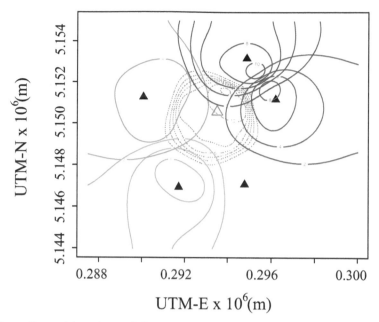

COLOR PLATE 4 (see chapter 4) Contour lines show the predicted home range pattern at Hanford ALE following removal of the Hopsage pack. The pattern of space use prior to removal are shown by the fitted probability density functions in color plate 3. The location of the former home range center of Hopsage pack (△) and the locations of the home range centers of the remaining packs (▲) are also shown.

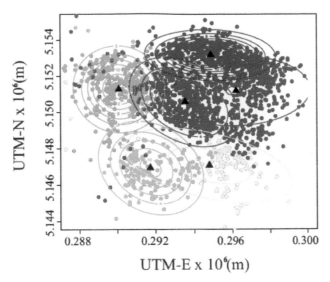

COLOR PLATE 5 (see chapter 4) Statistical home range model fit to the Hanford radio-location data using bivariate normal probability density functions for each pack.

(a)

(b)

Color Plate 6 (see chapter 5) Introduction of a new pack illustrates the interdependency in the spatial organization of home ranges. The surface height indicates the sum of expected space use by each group, while surface's color indicates the total density of scent marks. (a) With only four packs present, the home ranges are broad and overlapping, the boundaries between neighboring groups are only lightly scent-marked. (b) Adding a fifth pack into the central interior causes home ranges to become more distinct and non-overlapping. Sharp boundaries form, both between existing groups and between the new group and its neighbors.

(a)

(b)

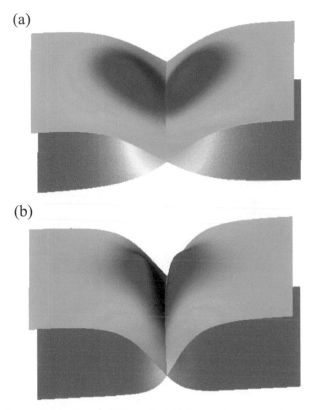

COLOR PLATE 7 (see chapter 5) The effect of changes in movement parameter β on the home range distributions for a pair of groups occupying a rectangular domain. At low values of β (panel a), the surfaces of expected space use for each pack decline gradually, reflecting a high degree of home range overlap. Scent marks are spread out, with a broad peak of high scent-mark density and a large area of intermediate scent-mark density (lighter shading at the intersection of the two surfaces). At higher values of β (panel b), expected space use decreases more sharply with distance from each home range center, indicating reduced home range overlap, and scent marks are concentrated at the edge of the home range (dark shading at the intersection of the two surfaces).

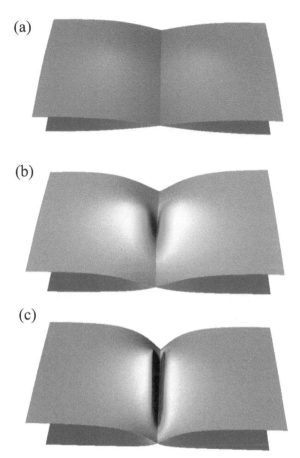

COLOR PLATE 8 (see chapter 5) Variation in the scent-marking response parameter m alters the spatial distribution of scent marks. (a) When $m = 0$, scent-mark densities are highest at the home range center. (b) Scent marking in response to the presence of foreign scent ($m > 0$) stimulates the formation of a scent-marked territorial boundary on the periphery of the home range. (c) As m increases further, the scent-marking boundary becomes sharper.

(a)

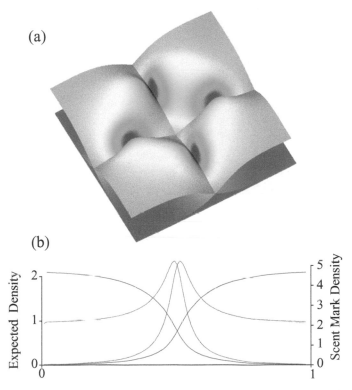

(b)

COLOR PLATE 9 (see chapter 5) Comparison of the two-dimensional and one-dimensional solution of the conspecific avoidance model. (a) shows the two-dimensional solution of equations (5.1)–(5.2) for four groups interacting on a square domain. The height of the surfaces indicates the expected space use by each group, while the shading reflects the density of scent marks (light (low)–dark (high)). (b) shows the solution of the analogous one-dimensional equations for a pairwise interaction on a linear domain. Panel shows the pattern of space use by each group (solid lines) and the spatial distribution of scent marks (dashed lines). In both one-dimensional and two-dimensional solutions, spatial segregation occurs; however, in two dimensions the scent marks form clusters between neighboring home ranges, a property not obvious from the corresponding one-dimensional solutions.

(a)

(b)

COLOR PLATE 10 (see chapter 7) Photographs illustrating the difference in terrain between (a) the Hanford Arid Lands Ecosystem (ALE) reserve and (b) Lamar Valley, Yellowstone National Park. The plume of smoke rising from the center of panel (a) is from the Hanford nuclear facility.

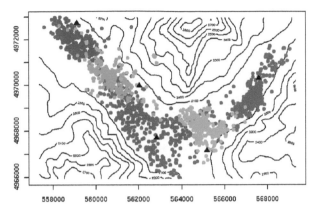

COLOR PLATE 11 (see chapter 7) Radio-tracking data for five contiguous packs in the Lamar Valley region of Yellowstone National Park collected by Crabtree et al. (unpublished data). Contour lines show topographic relief in meters. Colored points indicate radio locations for individuals belonging to the Bison (red points), Druid (green points), Fossil Forest (blue points), Norris (light blue), and Soda Butte (purple) packs collected during the period 1990–1994 prior to wolf reintroduction. ▲ indicate the typical denning areas for each pack used in the model fitting as their home range center.

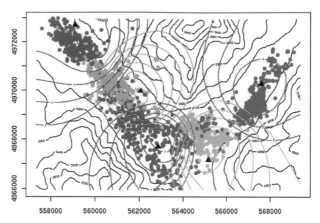

COLOR PLATE 12 (see chapter 7) Contour lines showing fit of the conspecific avoidance (CA) home range model (eqs. (4.20)–(4.21)) to relocation data (•) for the five groups in Lamar Valley collected by Crabtree et al. (unpublished data). Gray contour lines show topographic relief in meters. Colored contour lines show the probability density function $u^{(i)}(x, y)$ for each pack ($i = 1 \ldots 5$) plotted using a contour interval of 2, in density units scaled so that both the domain area A and integral of $u^{(i)}(x, y)$ are unity. The home range centers for each pack are shown (▲). Maximum log-likelihood values and estimates for β and m are given in table 7.1.

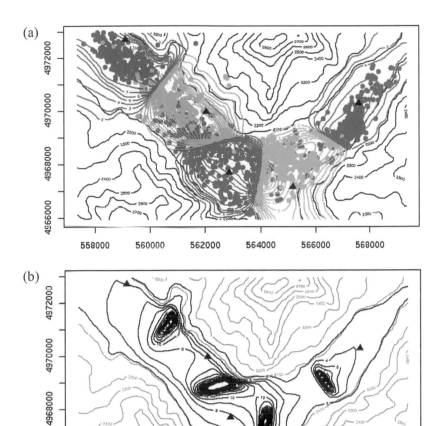

COLOR PLATE 13 (see chapter 7) Contour lines showing fit of the steep terrain avoidance plus conspecific avoidance (STA+CA) home range model (eqs. (7.6)–(7.7)) to relocation data (•) for the five coyote packs in Lamar Valley collected by Crabtree et al. (unpublished data). (a) The probability density function $u^{(i)}(x, y)$ for each of the five packs ($i = 1 \ldots 5$) plotted using a contour interval of 2, in density units scaled so that both the domain area A and integral of $u^{(i)}(x, y)$ are unity. The home range centers for each pack are shown (▲). (b) Contour lines showing the expected regional pattern of total scent-mark density $\sum_{i=1}^{5} p^{(i)}$ within Lamar study area, obtained from the fit of the STA+CA home range model. Total scent-mark density across the domain is shown using a contour interval of 2. Maximum likelihood values and estimates for β, m and α_z are given in table 7.1.

(a)

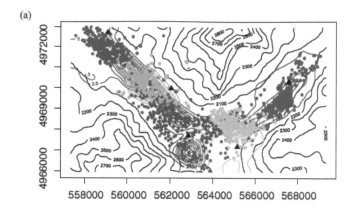

(b)

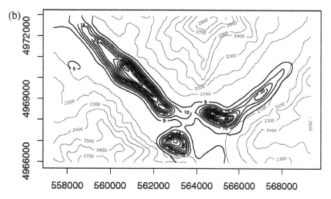

COLOR PLATE 14 (see chapter 7) Contour lines showing fit of the "conspecific avoidance plus prey availability" (PA+CA) home range model (eqs. (7.14) and (7.7)) to
relocation data (•) for the five coyote packs in Lamar Valley collected by Crabtree et
al. (unpublished data). (a) The probability density function $u^{(i)}(x, y)$ for each of the
five packs ($i = 1 \ldots 5$), in density units scaled so that both the domain area A and
integral of $u^{(i)}(x, y)$ are unity. The home range centers for each pack are shown (▲).
(b) Contour lines showing the expected regional pattern of total scent-mark density
$\sum_{i=1}^{5} p^{(i)}$ within Lamar study area, obtained from the fit of the PA+CA home range
model (eqs. (7.14) and (7.7)). Total scent-mark density across the domain is shown
using a contour interval of 2. Maximum log-likelihood values and estimates for β, m
and α_r are given in table 7.1.

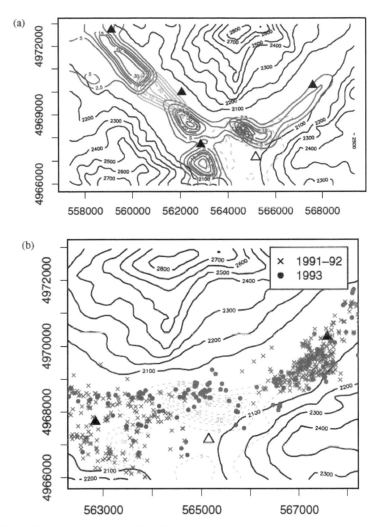

COLOR PLATE 15 (see chapter 7) (a) Colored contour lines show the spatial distribution of home ranges in Lamar Valley predicted by the PA+CA model following loss of the centrally located Norris pack in January 1993. Solid contour lines show the new configuration of home ranges in relation to the former home range of the Norris pack indicated by the dashed light-blue contour lines. Patterns of space use prior to the removal are shown in color plate 14b. Black contour lines show topographic relief in meters. The location of the former home range center of the Norris pack (△) and the locations of the home range centers of the remaining packs (▲) are also shown. (b) Observed changes in space use by the Soda Butte and Fossil Forest packs following the loss of the Norris pack in January 1993. As in (a), the dashed light-blue contour lines show the former home range of the Norris pack. Colored points show the relocations of individuals in the neighboring Soda Butte (purple) and Fossil Forest (blue) packs prior to (×) and following (•) the loss of the Norris pack As predicted by the model (panel a), the Soda Butte and Fossil Forest packs expand their territories into the former Norris home range following the loss of the Norris pack, focusing their movements on the prey-rich mesic grassland north of the former Norris home range center (△).

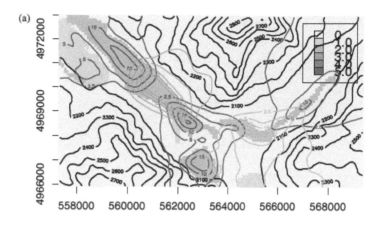

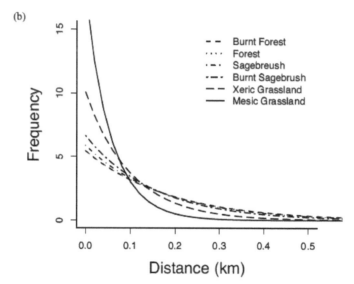

COLOR PLATE 16 (a) (see chapter 7) Colored contour lines show the predicted spatial distribution of home ranges in Lamar Valley following a 50% reduction in small mammal biomass. Gray shading shows the spatial pattern of prey availability in kg ha^{-1} following the reduction. Black contour lines show topographic relief in meters. The home range centers for each pack are also shown (▲). PLATE (b) Lines showing the expected distributions of movement distances for the different habitats within the Lamar valley study area obtained from the fit of the PA+CA model (eqs. (7.14) and (7.7)) shown in color plate 14a. Distributions were calculated using the relationship between α_r, resource availability $h(x, y)$, and the mean movement distance of individuals assuming that individuals are relocated every 5 minutes and that the mean movement speed of individuals in the absence of food is 2.9 km h^{-1} (i.e., $\rho_0 = 0.24$ km in eq. (7.13)).

TABLE 7.2. Estimates of small mammal density in the different habitats found in Lamar Valley and surrounding areas. Density estimates for each species were estimated from mini-grid trapping (Crabtree, unpublished data).

	Density (# per ha)			
Habitat	Mice (*Microtus*)	Ground Squirrels	Pocket Gophers	Red-backed Voles
Mesic Grass	79.76	0.0	15.02	0.0
Xeric Grassland	8.95	0.83	26.7	0.0
Sagebrush	8.13	0.17	8.48	0.0
Burned Sagebrush	4.47	1.82	10.8	0.0
Forest	0.0	0.0	8.53	8.33
Burned Forest	0.9	0.0	9.41	3.13
Mean weight (g)	50	250	100	20

are higher. Table 7.2 shows the density of small mammals associated with different habitat types found in Lamar Valley. There is a high degree of spatial variability in small-mammal densities in the different habitat types, in particular, mesic grassland areas contain high densities of mice (*Microtus spp.*) and pocket gophers (see table 7.2). Multiplying the density estimates for each species by the mean body weight for that species and summing the resulting numbers yields an estimate for the small mammal biomass associated with each habitat type. Combining these numbers with the Yellowstone National Park Geographic Information System habitat data layer, we can calculate the spatial distribution of total small-mammal biomass in Lamar Valley (figure 7.1).

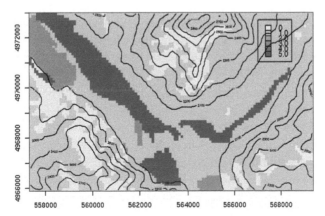

FIGURE 7.1. Small mammal densities in the Lamar Valley region of Yellowstone National Park collected by Crabtree et al. (unpubl. data). Shading indicates small mammal densities (kg ha^{-1}) calculated using the figures in table 7.2 mapped onto the habitats in the region using the Park Service Geographic Information System. Contour lines show topographic relief in meters.

As can be seen from figure 7.1, the patches of mesic grassland habitat found in the low-elevation areas surrounding the Lamar River have considerably higher densities of small mammals than higher elevation areas. Based on these observations, we now develop an alternative home range model formulation for the spatial distribution of relocations seen in color plate 11, in which individuals adjust their movement behavior in response to variation in small-mammal abundance.

Incorporating Foraging Responses to Resource Availability

Results from several studies have shown how resource availability affects the fine-scale foraging movements of carnivores. For example, an analysis of coyote movement patterns in Idaho using high-frequency radio tracking by Laundre and Keller (1981) showed that when foraging, individuals move slowly, at speeds between 0.5 and 1.5 km h^{-1}, making frequent turns (Type A and B movements, figure 7.2a, b). In contrast, when moving between foraging areas, individuals moved more quickly, at speeds between 1.5 and 4 km h^{-1}, and with fewer changes in direction (Type C movements, figure 7.2a, c). Macdonald (1980b) found similar effects of abundance on the foraging behavior of red foxes: in areas of high earthworm density, where prey capture rates were high, individuals turned frequently, resulting in convoluted movement paths, while in areas of low earthworm abundance, where prey capture rates were low, individuals turned less frequently, resulting in more directed movement.

The effects of resource availability on movement behavior is also seen in activity budgets, where it results in correlations between prey availability and time spent in different areas (Murray et al. 1994; Gese et al. 1996a; Gese et al. 1996b). For example, an analysis of coyotes foraging on small mammals in Yellowstone (Gese et al. 1996a; Gese et al. 1996b) found that differential habitat use was attributable to differences in foraging success, with an individual's rate of prey capture accounting for 84% of the variation in time spent foraging in different habitat types (figure 7.3).

In the STA+CA model, both differences in scent-mark density and terrain steepness induced changes in an individual's distribution of subsequent movement directions—so-called *tactic* movement responses. However, as described above, when foraging on small prey items, carnivores typically respond to changes in resource availability not by changing their distribution of movement directions but by changing their speed of movement and/or frequency of turning—so-called *kinetic* movement responses (Tranquillo and Alt 1990; Okubo and Grunbaum 2001). A further distinction is sometimes made between changes in movement speed (*orthokinesis*), and changes in turning frequency (*klinokinesis*) (Tranquillo and Alt 1990). As illustrated by

(a)

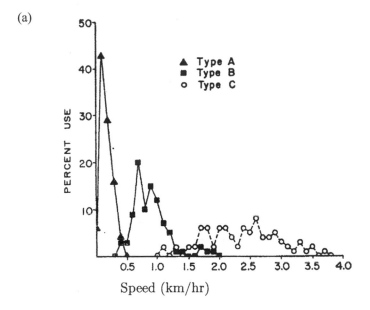

Speed (km/hr)

(b) Type A and B movements:

(c) Type C movements:

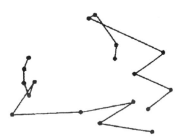

FIGURE 7.2. Measurements of coyote movement patterns using high-frequency radio tracking. (a) Distribution of movement speeds associated with foraging (Type A and B movements) and moving between foraging areas (Type C movements). (b) and (c). Patterns of movement associated with movement speeds shown in panel (a). Redrawn from Laundre and Keller (1981).

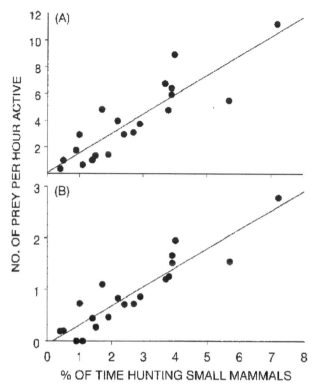

FIGURE 7.3. Relationship between prey capture rate and the amount of time coyotes spent hunting in different habital types. Redrawn from Gese et al. (1996a).

Okubo and Grunbaum (2001), both of these kinetic responses alter patterns of space use by changing the mean-squared displacement of an individual per unit time in different areas.

For example, suppose that consistent with the observations seen in figure 7.2, individuals exhibit a reduction in movement speed and turning frequency in response to increasing resource abundance. When considering the actual movement path of an individual, it is possible to distinguish between these two components of the individual's kinetic response to increasing resource availability. However, this distinction is not apparent in an implied movement trajectory calculated from relocations taken at a fixed sampling frequency. Due to the fixed time interval between relocations, both types of kinetic response reduce the distance the individual moves between successive relocations in areas of high resource density.

For simplicity, we consider the case where there is no steep terrain avoidance by the individual ($\psi = 1$ in eq. (7.5)), but the distance between relocations, ρ_τ (see eq. (7.5)), is a decreasing function of local resource density, i.e., $\rho_\tau(h(\mathbf{x}))$, where $h(\mathbf{x})$ is the resource density at location $\mathbf{x} = (x, y)$, shown in figure 7.1. Substituting $\rho_\tau(h(\mathbf{x}))$ into equations (7.5) with ψ set to 1, we obtain the following equations for expected space use:

$$0 = \underbrace{\nabla^2 \left[d(\mathbf{x}) u^{(i)} \right]}_{\text{foraging response}} - \underbrace{\nabla \cdot \left[c_P(\mathbf{x}) \vec{\mathbf{x}}_i u^{(i)} \sum_{j \neq i}^{n} p^{(j)} \right]}_{\text{scent-mark avoidance}}, \quad i = 1 \ldots n \quad (7.10)$$

where

$$d(\mathbf{x}) = \overline{\rho_\tau^2}(h(\mathbf{x}))/(4\tau) \quad \text{and} \quad c_P(\mathbf{x}) = b\overline{\rho_\tau^2}(h(\mathbf{x}))/(2\tau). \quad (7.11)$$

The first term in equation (7.10) can be expanded to yield

$$0 = \underbrace{\nabla \cdot \left[d(\mathbf{x}) \nabla u^{(i)} \right]}_{\text{random motion}} - \underbrace{\nabla \cdot \left[c_P(\mathbf{x}) \vec{\mathbf{x}}_i u^{(i)} \sum_{j \neq i}^{n} p^{(j)} \right]}_{\text{scent-mark avoidance}}$$

$$+ \underbrace{\nabla \cdot \left[u^{(i)} \nabla d \right]}_{\text{directed movement toward areas of high resource density}}, \quad i = 1 \ldots n$$

$$(7.12)$$

indicating that the increased turning frequency that occurs in response to increasing resource density results in a directed component of motion toward areas of high resource density where d($\mathbf{x}$) is low. The equations for scent marking (eqs. (7.7)) remain unchanged.

As noted in section 4.3, if the distribution of step lengths is exponential with mean $\bar{\rho}_\tau$ then $\overline{\rho_\tau^2} = 2\bar{\rho}_\tau^2$. Assuming that the relationship between mean step length $\bar{\rho}_\tau(h(\mathbf{x}))$ and prey density $h(\mathbf{x})$ is also exponential, i.e.,

$$\bar{\rho}_\tau(h(\mathbf{x})) = \sqrt{\tau}\rho_0 \exp(-\rho_1 h(\mathbf{x})) \quad (7.13)$$

(see figure 7.4). Substitution of equation (7.13) into the above definitions for $d(\mathbf{x})$ and $c_P(\mathbf{x})$ yields a simplified version of equation (7.12):

$$0 = \underbrace{\nabla \cdot \left[e^{-\alpha_r h(\mathbf{x})} \nabla u^{(i)} \right]}_{\text{random motion}} - \underbrace{\nabla \cdot \left[e^{-\alpha_r h(\mathbf{x})} c_P(\mathbf{x}) \vec{\mathbf{x}}_i u^{(i)} \sum_{j \neq i}^{n_{\text{pack}}} p^{(j)} \right]}_{\text{scent-mark avoidance}}$$

$$\underbrace{- \nabla \cdot \left[e^{-\alpha_r h(\mathbf{x})} u^{(i)} \nabla h \right]}_{\text{directed movement toward areas of high resource density}}, \quad (7.14)$$

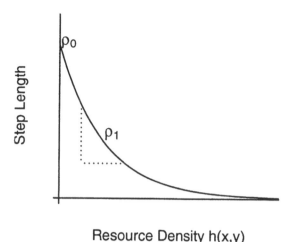

Resource Density h(x,y)

FIGURE 7.4. Relationship between mean step length $\bar{\rho}_\tau$ and resource density in the "prey availability plus conspecific avoidance" (PA+CA) model (eq. (7.14)). ρ_0 is an individual's step size in the absence of resources and ρ_1 determines the rate which the individual's mean step length decreases with increasing resource availability $h(x,y)$.

where $\beta = b$ and $\alpha_r = 2\rho_1$. The effect of prey density on movement behavior is reflected in the last term of the equation, which is a classic "prey taxis" term (Kareiva and Odell (1987); see also White et al. (1996)).

Equation (7.14) is subject to zero-flux boundary conditions and the scent-marking equations for individuals are unchanged (eq. (7.7)). We refer to equation (7.14) plus equation (7.7) as the "prey availability plus conspecific avoidance" (PA+CA) model.

Color plate 14 shows solution of the PA+CA model for the Lamar Valley coyote packs. As can be seen in the figure, the reduced movement distance per unit time in areas of high resource density changes the patterns of home ranges and scent marks, shifting the expected patterns of space for the Lamar packs toward areas of high resource density (compare color plate 14 figure 7.1). Comparison of the fit obtained with the PA+CA model to that obtained with the STA+CA model (compare color plates 14a and 13a) shows that incorporating a foraging response to small mammal density gives a substantially improved fit to the relocations than incorporating steep terrain avoidance (ΔAIC = 913.8, see table 7.1). For discussion of AIC, see chapter 4.

Several factors contribute to the improved goodness-of-fit obtained with the PA+CA model. First, the home ranges of the Bison, Druid, and Soda Butte packs (indicated by the red, green, and purple contour lines respectively) are narrower than those predicted by the STA+CA model, which gives a better fit to the spatial distributions of relocations for these three packs (compare color plates 14a and 13a). In addition, the PA+CA model accurately captures the partitioning of space between the Druid and Fossil Forest packs (green and blue contour lines respectively). While the STA+CA model predicts box-shaped home ranges that are approximately equal in size for the Druid and Fossil Forest packs, the PA+CA model predicts an elongated home range for the Druid pack and a larger sized home range for the Fossil Forest pack, which more closely matches the observed spatial distribution of relocations for these two groups (compare color plates 14a and 13a). Finally, the PA+CA model more accurately captures the cluster of Norris relocations (light blue points and contour lines) around the high resource area located shortly after the valley changes from running in a southeast direction to a northeast direction (compare color plate 14a and figure 7.1).

The qualitative spatial distribution of scent marks predicted by the PA+CA model is similar to that predicted by the STA+CA model (compare color plates 14b and 13b), however, the overall density of marks predicted by the PA+CA model is lower, reflecting the reduced overmarking response parameter (see table 7.1).

7.3. MODEL PREDICTIONS

The mechanistic nature of the models used in the above analysis means that the model fits can be used to predict how patterns of space use will change in response to perturbation. For example, in January 1993 following the death of the alpha male of the Norris pack, whose home range is indicated by the light-blue contour lines in color plate 14a, dissolved. Color plate 15a shows the changes in home-range configuration that the PA+CA model fit predicts will occur in response to this demographic perturbation. Individuals of the neighboring Soda Butte pack (purple contour lines) move into the former Norris home range, focusing their movements around the prey-rich mesic grassland situated north of the former Norris den site. The increased utilization of this high-resource area by the Soda Butte pack is accompanied by a contraction of their home range around the ribbon of mesic grassland habitat running along the valley floor, as the individuals focus their movements around these resource-rich areas (compare the purple contour lines in color plates 14a and 15a). Individuals of the neighboring Fossil Forest pack (blue contour lines) also move into the former Norris home range, again focusing their movements on the mesic grassland north of the former Norris den site. However, their shift in space use is less marked than that of the Soda Butte pack (compare the blue contour lines in color plates 14a and 15a).

These predicted changes in spatial configuration are consistent with the changes in patterns of space use seen in 1993 following the loss of the Norris pack. Color plate 15b shows the spatial pattern of relocations of the Soda Butte and Fossil Forest packs observed during 1991–1992 (×) and 1993 (•). As predicted, the Soda Butte and Fossil Forest packs move into the areas occupied by the former Norris home range, focusing their movements on the high-resource mesic grassland area north of the former Norris den site, and, as predicted, the individuals of the Soda Butte pack also focus their movements around the mesic grassland habitat on the valley floor (compare color plates 15a and 15b). Comparing the patterns of space use predicted before (color plate 14a) and after (color plate 15a) the loss of the Norris pack to the spatial distribution of relocations observed in 1993 shows a marked improvement in the model's goodness-of-fit (ΔAIC = 39.6), indicating that the model correctly predicts the changes in space use that occurred following the loss of the Norris pack (see table 7.3).

In a similar manner, color plate 16a shows the predicted consequences of a 50% decrease in food availability for patterns of space use in Lamar Valley obtained from the PA+CA model fit. While no detailed quantitative data are available, the reduction in food availability causes individuals to move more extensively, resulting in larger home ranges and increased home range overlap between packs, a response that is qualitatively consistent with seasonal changes in home range patterns observed in coyote populations (Bekoff and Wells 1986).

In addition to these large-scale predictions, the model fits also yield predictions for how the fine-scale movement behavior of individuals varies across the landscape. For example, we can use the estimate of α_r from the PA+CA model fit in conjunction with equations (7.13)–(7.14), and the estimates of prey availability in the different habitats (table 7.2), to calculate the predicted distribution of distances between relocations in different habitat types. Color plate 16b shows the expected distribution of movement distances in different habitat types for individuals relocated at five-minute intervals, assuming that their average movement speed in the absence of food availability is 2.9 km h^{-1}—a number consistent with Laundre and Keller's (1981) observations of coyote movement behavior described earlier. As a result of high prey availability, the distance between successive relocations in the mesic grasslands, and to a lesser

TABLE 7.3. Goodness-of-fit of the PA+CA model to the 1993 relocations before and after the removal of the Norris pack. The total number of relocations in 1993 was 456. The significant improvement in the fit of the model following the removal of the Norris pack indicates that the PA+CA model correctly predicts the changes in space use that occurred after the Norris pack dissolved in January 1993.

	Before Norris removal	After Norris removal
$l(\theta)$	747.4	767.2
AIC	−1488.8	−1528.4

extent in the xeric grasslands, are predicted to be substantially lower than in the forested and sagebrush habitats (color plate 16a and plate 16b). These predictions are consistent with independent observations of coyote behavior in Lamar Valley (Gese et al. 1996a; Gese et al. 1996b).

7.4. SUMMARY

This chapter illustrates how mechanistic home range models incorporating multiple movement cues can be used to determine the underlying ecological causes of home range patterns, and predict observed changes in space use in response to perturbation. Specifically, terrain heterogeneity and prey availability were evaluated as two potential explanations for the observed spatial distribution of coyote home ranges in Lamar Valley, Yellowstone. By incorporating movement responses to these two sources of heterogeneity into correlated random walk models of individual movement behavior, and then evaluating the resulting predictions for patterns of space use against the observed spatial distribution of relocations, it was possible to show that heterogeneity in prey availability provides a better explanation for the observed patterns of coyote space use than avoidance of steep terrain. The fitted model incorporating responses to prey availability correctly predicts the changes in space use that occurred following the loss of one of the packs from the study region in 1993.

The ability of mechanistic home range models to predict patterns of space use for individuals that are responding to multiple orientation cues removes an important barrier to their application in many empirical situations where the movements of individuals are being influenced by a variety of different physical and biological factors as they traverse heterogeneous landscapes. As illustrated here, by incorporating different movement terms and comparing the goodness-of-fit to observed patterns of space use, mechansitic home range models can be used to determine the underlying causes of observed patterns of space use in natural populations. In addition to the influences of terrain and resource availability considered here, patterns of space use in some carnivore populations are also affected by the spatial distribution of competitors, an issue that we will explore further in chapter 9.

CHAPTER EIGHT

Home Range Formation in the Absence of a Den Site

In the mechanistic home range models discussed so far, the den site has acted as a focal point for the movement of individuals. In this chapter, we briefly explore an alternative mechanism for the process of home range formation in the absence of home range center.

In some populations of carnivores, home ranges have been observed to form in the absence of surrounding packs (Mech 1991) and in the absence of a den site (Mech, personal communication, 1994). One possible explanation is simply that, rather than a den site, a core foraging area or rendezvous site is acting as the focal point for the movement of individuals. If this were the case, the localizing tendency and conspecific avoidance home range models discussed in the earlier chapters would also be applicable to these situations, with the coordinates of these core areas or rendezvous sites substituted as the focal point for directed movement.

However, during the early stages of home range establishment following pair formation, newly formed pairs move together and engage in high levels of scent marking and overmarking of each others' scent marks (Rothman and Mech 1979; MacDonald 1979). Based on these observations, Briscoe et al. (2002) proposed a home range model in which individuals exhibit an overmarking response to familiar scent marks and preferentially remain in areas of high scent-mark density.

8.1. MODEL FORMULATION

Suppose that a newly formed pack, consisting of a pair of individuals, has no fixed den site. In the absence of encounters with familiar scent marks, the pair move together randomly according to equation (2.26). However, encounters with familiar scent marks reduce their propensity to move, causing them to preferentially remain in areas with a high density of familiar scent marks. Specifically, the Briscoe et al. (2002) home range model assumes that an

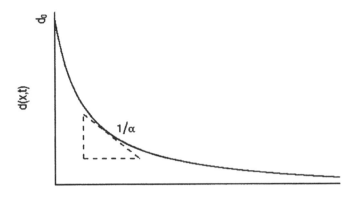

FIGURE 8.1. Relationship between mean-squared displacement and the density of familiar scent marks (eq. (8.1)).

individual's probability of moving at each time step is a decreasing function of the density of familiar scent marks encountered. Using the methods described in chapter 2, it can be shown that this movement rule does not give rise to any directed movement but instead causes the mean-squared displacement per unit time of the pair to decrease monotonically with the density of familiar scent marks encountered (figure 8.1).

As a result of the above relationship, the diffusion coefficient, $d(\mathbf{x}, t)$ in equation (2.26), changes from being a constant d_0, to

$$d(\mathbf{x}, t) = \frac{d_0 \alpha}{\alpha + p(\mathbf{x}, t)} \tag{8.1}$$

where $p(\mathbf{x}, t)$ is the density of scent marks encountered, and the parameter α determines the rate at which the mean-squared displacement of the pair declines with the increasing $p(\mathbf{x}, t)$.

Substituting equation (8.1) into equation (2.26) yields the following equation for the pattern of space use by the pair:

$$\frac{\partial u}{\partial t} = \nabla^2 \left(\frac{d_0 \alpha u}{\alpha + p} \right) \tag{8.2}$$

In addition to affecting movement behavior, encounters with familiar scent marks also change the rate of scent marking. Suppose that in the absence of familiar scent marks, the pair scent-mark at rate l, and their scent-marking rate increases as a function of the density of familiar marks encountered. The rate of change of scent-mark density at each location $\mathbf{x}$ will then be given by

$$\frac{dp(\mathbf{x}, t)}{dt} = u(\mathbf{x}, t) \left[l + M(p(\mathbf{x}, t)) \right] - \mu p(\mathbf{x}, t), \tag{8.3}$$

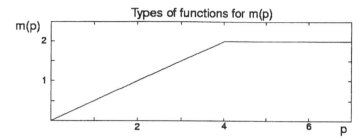

FIGURE 8.2. This figure shows the piecewise linear scent-response function, $M(p)$. Here $m = 0.5$ and $m_{max} = 2$.

where $u(\mathbf{x}, t)$ is the pattern of space use by the pair, μ is the rate at which marks decay, and the function $M(p(\mathbf{x}, t))$ governs the effect of familar scent marks on their rate of marking. The Briscoe et al. (2002) model assumes the following form for $M(p)$:

$$M(p) = \begin{cases} mp & 0 \leq p \leq m_{max} \\ m_{max} & p \geq m_{max} \end{cases}, \tag{8.4}$$

i.e., the pair's rate of scent marking increases linearly with the density of familiar scent marks encountered up to a level m_{max}, at which point the rate of scent marking does not increase further (figure 8.2).

8.2. ANALYSIS

As in the previous chapters, we begin our analysis by nondimensionaliz-ing equations (8.2)–(8.4) to reduce the number of parameters. Defining the following nondimensional variables:

$$x^* = \frac{x}{L}, \quad y^* = \frac{y}{L}, \quad t^* = \mu t, \quad u^* = uL^2, \quad p^* = \frac{L^2 \mu p}{l},$$

$$d_0^* = \frac{L^2 d_0}{\mu}, \quad \alpha^* = \frac{\alpha l}{\mu L^2}, \quad m^* = \frac{m}{\mu L^2}, \quad m_{max}^* = \frac{m_{max}}{l}, \tag{8.5}$$

then making the above substitutions into equations (8.2)–(8.4) and dropping the asterisks, we get:

$$\frac{\partial u}{\partial t} = \nabla^2 \left(\frac{d_0 u}{1 + p} \right) \tag{8.6}$$

and

$$\frac{dp(\mathbf{x}, t)}{dt} = u(\mathbf{x}, t) \left[1 + M(p(\mathbf{x}, t)) \right] - p(\mathbf{x}, t). \tag{8.7}$$

Note that the scent-mark response function $M(p, t)$ in equation (8.7) is the same as equation (8.4) but with m and m_{max} now in nondimensional form. Distances (x and y) are now scaled to the size of the area under consideration ($L = A^{1/2}$, where A is the area of the study region), time (t) is scaled to the rate of scent-mark decay (μ), and scent-mark density (p) is scaled in relation to the individual's rate of marking in the absence of familiar scent marks (l).

Figure 8.3 shows numerical solutions of equations (8.6)–(8.7) in a single space dimension. As can be seen from the figure, the pack develops a well-defined home range despite the absence of a defined home range center, and in the absence of any neighboring packs. The general shape of the home range predicted by this model is reminiscent of the home ranges arising from the conspecific avoidance model (see figure 4.2 and color plate 3), consisting of a "flat-topped" region within which space use is relatively uniform, with abrupt edges, outside of which space use by the pack is near zero.

While the patterns of space use are similar, the mechanisms underlying home range formation in the two models are fundamentally different. In the conspecific avoidance model, home ranges arise from encounters with foreign scent marks, generating an avoidance response directed toward a defined home range center, while in the above model, home ranges arise from individuals exhibiting a propensity to remain in areas of familiar scent marks.

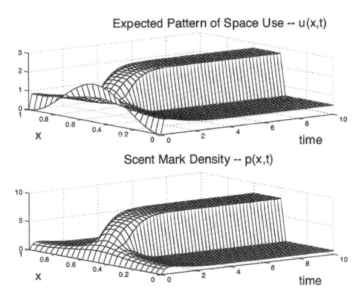

FIGURE 8.3. Numerical solutions of equations (8.6) and (8.7) in a single space dimension (x). The pair develops a distinct home range and distribution of scent marks despite the absence of a defined home range center.

Further mathematical analysis of the above model indicates that under certain initial conditions, equations (8.6)–(8.7) do not give well-behaved solutions, as a result of the intrinsic positive feedback between increased space use and increased scent marking (Briscoe et al. 2002). This undesirable behavior can be eliminated by introducing a small amount of spatial averaging into equation (8.7), implying that individuals are responding to the average density of scent marks in a small region around their current position (see Briscoe et al. 2002).

The lack of an avoidance response to foreign scent marks in this alternative home range model formulation means that the model is most applicable to newly formed pairs colonizing unoccupied areas. In established populations, avoidance responses are likely to be more important in governing patterns of space use. However, it is possible that even in these circumstances, individuals may also localize their movement in areas containing familiar scent marks, particularly during the process of home range formation when a den site has not yet been established. A modified version of this model features in our exploration of interspecific interactions in the latter half of the next chapter.

8.3. SUMMARY

In both the localizing tendency and conspecific avoidance home range models, a den site or core area acts as a focal point for the movement of individuals. This chapter explores an alternative home range model formulation in which individuals increase their rate of scent marking in the presence of familiar scent marks and reduce their rate of movement in areas of existing scent marks. Numerical simulations of this model show that these movement rules can give rise to "flat-topped" home ranges with abrupt edges in the absence of a den site or core area. This model is appropriate for the early stages of home range formation and for individuals entering new areas where avoidance responses are likely to be unimportant due to minimal contact with other packs.

CHAPTER NINE

Secondary Ecological Interactions

In this chapter, we consider two extensions to the home range models analyzed in the earlier chapters that explore the effects of space use on spatial interactions with prey populations and competing carnivore populations. As examples of such secondary ecological interactions, we consider the interaction between wolves and white-tailed deer in northeastern Minnesota, and the interaction between wolves and coyotes in areas where these two canids co-occur such as Yellowstone National Park.

9.1. WOLF–DEER INTERACTIONS

In northeastern Minnesota, the main prey species for wolves is the white-tailed deer (*Odocoileus virginianus*), which accounts for approximately two-thirds of wolf food intake (Van Ballenberghe 1972; Fritts and Mech 1981). Wolf predation is a major factor affecting white-tailed deer population across the region, accounting for 90% of all known deaths (Nelson and Mech 1981).

Deer are primarily found on the periphery of wolf home ranges and in the "buffer zones" between the adjacent pack territories (figure 9.1). Their spatial distribution within these areas varies seasonally: in the summer months the deer disperse onto individual home ranges, but in winter they congregate in "deer yards" where they trample the snow and establish a network of trails (Nelson and Mech 1981).

This spatial patterning in the distribution of deer arises despite a relatively uniform distribution of forage, and despite the fact that deer do not avoid settling in regions of high wolf density. This implies that the negative correlation between deer locations and wolf home ranges arises as a result of differential predation rates between the interior of wolf home ranges and the buffer zones that separate them. This led Nelson and Mech (1981) to suggest that the buffer zones between wolf territories may act as a "prey refuge" for deer (see also Taylor and Pekins 1991).

We explore the spatial dynamics of this interaction using a simple model in which a pair of wolf packs, moving and interacting on a symmetric

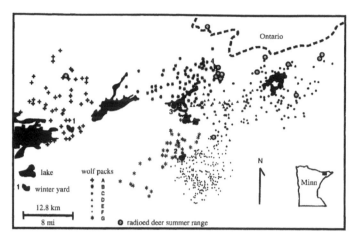

FIGURE 9.1. Winter yards and summer ranges of radio-collared deer in relation to wolf
pack territories (based on Hoskinson and Mech 1976).

one-dimensional domain, prey on a deer population within the region. Our
analysis closely follows that of Lewis and Murray (1993), which assumes that
the functional response of the wolves to local deer density is linear. Therefore,
the spatial distribution of the deer population with the region, $h(x)$, is given by:

$$\frac{\partial h(x)}{\partial t} = \underbrace{-\psi (u(x) + v(x))h(x)}_{\text{predation}} - \underbrace{\mu_h h(x)}_{\text{mortality}} , \qquad (9.1)$$

where $h(x)$ is the deer density, $u(x)$ and $v(x)$ are the expected patterns of space
use by the two packs, ψ is the predation rate, and μ_h is the natural mortality
rate of deer in the absence of predation.

The production of deer each spring is assumed to be given by the Beverton-
Holt density-dependent recruitment equation:

$$h(T^+, x) = \frac{\lambda h(T^-, x)}{1 + (\lambda - 1)h(T^-, x)/K} \qquad (9.2)$$

where $h(T^-, x)$ and $h(T^+, x)$ are the local density of deer immediately prior to
and after reproduction during the spring, and λ and K are the growth rate and
carrying capacity of the deer in the absence of predation.

During the year, predation by wolves according to equation (9.1) reduces
deer numbers such that the deer population immediately prior to the spring birth
in the following year is

$$h((T + 1)^-, x) = h(T^+, x) \exp(-\psi [u(x) + v(x)] - \mu_h). \qquad (9.3)$$

A "steady-state" solution, which gives no change in density from spring to spring, satisfies $h((T+1)^+, x) = h(T^+, x)$. Substituting equation (9.1) and equation (9.2) into equation (9.3) and satisfying this condition yields the spring deer density immediately after birth as

$$h(T^+, x) = \max \left\{ 0, \frac{\lambda - \exp(\psi[u(x) + v(x)] + \mu_h)}{\lambda - 1} \right\}, \qquad (9.4)$$

where h is now the deer density relative to the carrying capacity K. Integrating equation (9.4) from T to $T+1$ yields the average deer density $H(u(x), v(x))$ throughout the year:

$$H(u(x), v(x)) = \max \left\{ 0, \frac{(\lambda - \exp([u(x) + v(x)]\psi + \mu_h))}{\lambda - 1} \right.$$
$$\left. \frac{(1 - \exp(-\psi(u(x) + v(x)) + \mu_h))}{\psi(u(x) + v(x))} \right\}. \qquad (9.5)$$

Examination of the above equation indicates that to maintain a deer population at a particular location x requires that the maximum number of offspring per adult deer per year, λ, exceed the local mortality rate $\exp([u(x)+v(x)]\psi + \mu_h)$.

If space use by wolves is spatially uniform, the deer population will not persist if the density of wolves exceeds $(\log(\lambda) - \mu_h)/\psi$. As can be seen from this expression, this is most likely when either deer recruitment is low $(\log(\lambda) - \mu_h)$ small or predation rates are high (ψ large). However, under the same circumstances, if the predation pressure is significantly non-uniform, then the deer population may persist in the areas of reduced predation pressure. Figure 9.2 shows solutions to equations (9.4) and (9.5) when the movement behavior of individuals in the two wolf packs is given by the conspecific avoidance home range model (equations (6.1)–(6.4)). For the parameter values shown in the figure, in the absence of the spatial patterning in wolf density, the deer population would go extinct. However, as can be seen in the figure, the buffer zones between the adjacent wolf home ranges arising from conspecific avoidance act as prey refuges that allow the deer population to persist, as suggested by Nelson and Mech (1981).

Thus during periods when deer recruitment is low or predation rates are high, the buffer zones between adjacent wolf home ranges provide a benefit to the packs by facilitating the persistence of their primary prey species. The question then immediately arises as to whether the presence of buffer zones is evolutionarily stable: since the benefits of increased prey are shared by both packs, the buffer zones could be susceptible to "raiding" by either pack. However, offsetting the benefits of raiding the buffer zone is the potential for aggressive encounters with the members of the neighboring group. This issue of the

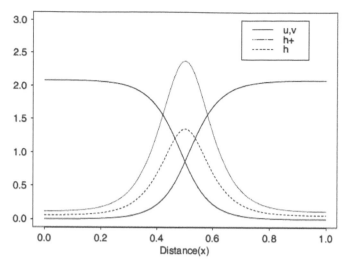

FIGURE 9.2. The region between wolf territories can act as a deer refuge. $u(x)$ and $v(x)$ are the patterns of space use by the two wolf packs, $h^+(x)$ is the deer density immediately after spring birth, and $h^-(x)$ represents the deer density immediately prior to spring birth. The wolf home ranges $u(x)$ and $v(x)$ are given by solutions of equations (6.1)–(6.4) with parameters $c/d = 7.34$ and $m = 0.48$. The "steady-state" yearly deer densities prior to reproduction and after reproduction are given by equations (9.3) and (9.4) respectively. Parameters for the deer dynamics were chosen as $\lambda = 2$, $\psi = 0.33$, and $\mu_h = 0$. For these parameter values, spatially uniform space use by the two packs (i.e., $u(x) + v(x) = 1$) causes the deer population to go extinct. Note that to facilitate comparison, the nondimensionalized deer density has been scaled by a factor of 10. In other words, the peak density at approximately 2.5 means the highest deer density is 25% of its carrying capacity. The presence of a buffer zone in the middle has a dramatic influence on the local deer levels. Here the total local density of wolves remains less than the critical value that can drive the deer extinct: $(\log(\lambda) - \mu_h)/\psi = 2.1$.

competitive stability of different movement strategies is considered further in chapter 11.

9.2. WOLF–COYOTE INTERACTIONS

In addition to their interactions with prey, wolves also exhibit a diverse array of interactions with a number of competitors, including bears, cougars, wolverines, foxes, and coyotes (Ballard et al. 2003). In this section we focus on the interaction between wolves and their closest competitor: coyotes. There is clear evidence that wolves have a significant negative impact on coyote populations. For example, since the reintroduction of wolves into Yellowstone National Park, 25–30% of the coyote population in Lamar Valley (see figure 7.2) has been killed by wolves, and overall coyote numbers have decreased by 50%

(Crabtree and Sheldon 1999). This phenomenon is widespread amongst canids and other carnivores (see Johnson et al. (1996) and Crabtree and Sheldon (1999) for reviews).

In this section we formulate a simple model for coyote space use in response to newly formed wolf territories. We assume that there is an initial overlap between the wolf territory and coyote home ranges, and that the coyote abandons its den site before searching for a new home range location. Studies by Paquet (1991) in Manitoba suggest that while wolves do not respond significantly to coyote scent marking and signs, coyotes do respond to wolf scent marking and signs. Based on these observations, we assume a unidirectional response of coyotes to wolf space use, rather than a two-way interaction.

Here we develop a simple mechanistic model for the spatial distribution of coyotes in which coyotes adjust their movements in response to wolf space use. Suppose that, as in the previous section, individuals belonging to a pair of wolf packs move and interact according to the conspecific avoidance home range model on a symmetric, one-dimensional domain (equations (6.1)–(6.4)). Suppose further that coyotes also occupy this region and their movement behavior is described by a variant of the movement model described in chapter 8, in which encounters with familiar scent marks decrease an individual's probability of movement. However, in this case, rather than a decreasing probability of movement in response to familiar scent marks, we specify that the probability of a coyote moving at each time step is an increasing function of wolf density. Using the same approach used in chapter 8, it is possible to show that the above movement rule results in the diffusion coefficient $d(x, t)$ coyotes being an increasing function of wolf density:

$$d(x, t) = \frac{d_0 w(x)}{\beta_{avoid} + w(x)}, \tag{9.6}$$

where β_{avoid} governs the sensitivity of coyote movement to wolf density and $w(x)$ is the spatial distribution of space use by wolves across the region. Although direct interactions between wolves and coyotes are rare, we assume that the coyotes are able to assess wolf space usage.

Substituting equation (9.6) into equation (2.26) yields the following equation for the pattern of coyote space use:

$$\frac{\partial s}{\partial t} = \frac{\partial^2}{\partial x^2} \left(\frac{d_0 w(x)}{\beta_{avoid} + w(x)} s(x) \right). \tag{9.7}$$

Assuming coyotes do not leave the area, then zero-flux boundary conditions are appropriate for equation (9.7). For simplicity, we focus on a single coyote, and thus the area under the curve $s(x)$ is 1.

We now calculate the pattern of coyote space use that arises in response to patterns of wolf space use. For the simple pairwise interaction considered here, total wolf space use $w(x) = u(x) + v(x)$, where $u(x) + v(x)$ are given by the solution of equations (6.1)–(6.4).

Steady-state solutions to equation (9.7) with zero-flux boundary conditions are found by setting the time derivative to zero, integrating once with respect to x, applying the boundary conditions, and dividing through by the constant d_0 to yield

$$\frac{\partial}{\partial x}\left[\frac{w(x)}{\beta_{avoid} + w(x)}s(x)\right] = 0. \tag{9.8}$$

One further integration with respect to x yields

$$s(x) = k(1 + \beta_{avoid}/w(x)). \tag{9.9}$$

Because the coyote is assumed not to leave the region, the area under the curve $s(x, t)$ is unity. This implicitly defines the constant k by

$$k = \int_\Omega \frac{w(x)}{\beta_{avoid} + w(x)}\, dx, \tag{9.10}$$

where Ω is the specified region.

Inspection of equation (9.9) indicates that if coyote sensitivity to wolves is small ($\beta_{avoid} \rightarrow 0$ in equation (9.9)), the coyote density becomes spatially uniform and is independent of wolf density. However, if the sensitivity of coyotes to wolves is large ($\beta_{avoid} \rightarrow \infty$), coyote density becomes inversely related to wolf density and the two species segregate spatially. This result is illustrated graphically in figure 9.3. The "coyote avoid" case illustrates a solution of equation (9.9) in which coyotes exhibit a high degree of sensitivity to wolves, and the "coyote coexist" case illustrates a solution to equation (9.9) in which coyotes have a low sensitivity to the presence of wolves.

This allopatric pattern of wolf and coyote space use that develops when coyotes exhibit a marked avoidance response to the presence of wolves is consistent with observed patterns of space use in sympatric populations of these two carnivores. According to Crabtree and Sheldon (1999), essentially all studies that have concurrently radio-tracked two different species of radio-tagged canids have indicated some degree of home range interspersion or fine-scale allopatry in their patterns of space use.

An interesting extension to the above model would be to incorporate the influence of space use on the demography of the two populations and determine the conditions under which a competitively inferior species is able to coexist with its competitive superior. In the case of wolves and coyotes, a more realistic model would also incorporate responses to coyote scent marks into the underlying model of coyote movement behavior.

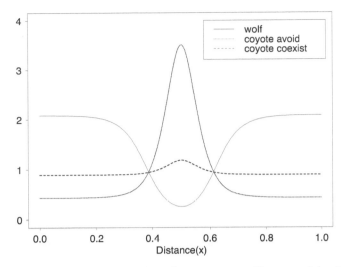

FIGURE 9.3. The shift in coyote territory in response to wolf scent mark levels. The "coyote avoid" case has $\beta_{avoid} = 1000$. The "coyote coexist" case has $\beta_{avoid} = 0.1$ (see eq. (9.9)).

9.3. SUMMARY

Evidence from field studies suggests that the spatial pattern of carnivore home ranges can have a significant impact on the spatial pattern of prey and competitors. In this chapter, we explore two simple extensions to the conspecific avoidance model that incorporate these secondary ecological interactions. In the first, we explore how the spatial pattern of home ranges arising from the avoidance of conspecifics can have a "top-down" effect on prey species. The results illustrate how spatial avoidance in carnivore populations gives rise to spatially variable predation rates that result in prey becoming concentrated at the boundaries between neighboring groups. During periods when prey recruitment is low or predation rates are high, these areas of high prey density act as reservoirs, allowing the prey population to persist under conditions when, in the absence of such spatial patterning, the population would otherwise go extinct. In the second, we explore spatial interactions between competing carnivore populations, developing a simple model of spatial avoidance by a competitively inferior species. The results from this model illustrate how spatial interactions can account for the fine-scale pattern of allopatric space use that is commonly observed to develop in populations of competing carnivores.

CHAPTER TEN

Displacement Distances:
Theory and Applications

As illustrated in figure 3.1, measurements of space use in carnivores typically show a rapid initial increase followed by a leveling off in the area used by the individual. In this chapter, we investigate the relationship between these kinds of empirical estimates of space use and the patterns of space use predicted by mechanistic home range models. We begin by analyzing the statistical properties of the minimum convex polygon (MCP) method widely used in empirical studies of animal home ranges, which utilizes the outermost relocations of an individual to produce an estimate of the individual's home range. We then turn our attention to two alternate statistics used in the analysis of correlated random walks that relate to the average space use by the individual, the mean-absolute and mean-squared displacement.

10.1. THE MINIMUM CONVEX POLYGON METHOD

As we noted in the introduction, one of the commonest methods for characterizing animal home ranges is the minimum convex polygon (MCP) method. An example of this method is shown in figure 10.2a. As can be seen in the figure, MCP estimates of an individual's home range are calculated by joining the outermost relocations to form a polygon whose internal angles are all less than 180 degrees. When using the MCP method, it is not uncommon for researchers to exclude the outermost 5% or 10% of observations prior to fitting the polygon to the data (e.g. Bowen 1982; Girard et al. 2002). The rationale for calculating such "trimmed" MCPs is that the outermost points tend to skew the area encompassed by the polygon, resulting in an overestimate for the home range size of the individual. As we show below, mathematical analysis of a simple mechanistic home range model establishes a mechanistic basis for this trimming procedure.

We examine the statistical properties of the MCP by considering the spatial distribution of furthest relocations that arise from an individual whose patterns of movement are governed by the Holgate-Okubo localizing tendency model analyzed in chapter 3. While our analysis in chapter 3 showed that this model's assumption of a fixed-magnitude localizing tendency is idealized, the model's simplicity makes it ideal for examining the statistical properties of displacement distances. Specifically, we ask: if we make repeated independent relocations of an individual whose movement behavior is governed by the localizing tendency model, what is the probability that the furthest relocation from the home range center is a given distance away, and how does this probability change with the number of relocations obtained?

Note that here the term "independent relocations" means that enough time elapses between successive relocations for the individual to reposition itself within its home range so its spatial relocation is independent of where it was at the previous time. This is generally a stricter criterion than the one used in chapter 2, which required only that the directions implied by successive relocations are uncorrelated. For procedures for calculating the time for the spatial position of successive relocations to become independent, see Swihart and Slade (1985, 1997) and Rooney et al. (1998).

We begin by considering the pattern of space use in a single dimension. As we showed in chapter 3, the pattern of one-dimensional space use $u(x, t)$ that arises from the Holgate-Okubo localizing tendency model is given by

$$\frac{\partial u}{\partial t} = \frac{\partial}{\partial x}(c \ \text{sgn}(x)u) + d\frac{\partial^2 u}{\partial x^2} \tag{10.1}$$

whose steady-state solution is

$$u(x) = \frac{\beta}{2} \exp(-\beta|x|), \tag{10.2}$$

where $\beta = c/d$.

Thus, the probability that relocation falls within distance s of the home range center is $1 - \exp(-\beta s)$. If n_r independent relocations are taken, then the probability that the furthest of n_r relocations falls within this distance is

$$P_{n_r}(s) = (1 - \exp(-\beta s))^{n_r}. \tag{10.3}$$

The probability density function for the distance of the furthest relocation $p_{n_r}(s)$ is given by the derivative of the above equation:

$$p_{n_r}(s) = \frac{dP_{n_r}(s)}{ds} = \beta n_r \exp(-\beta s)(1 - \exp(-\beta s))^{(n_r-1)} \tag{10.4}$$

(see figure 10.1).

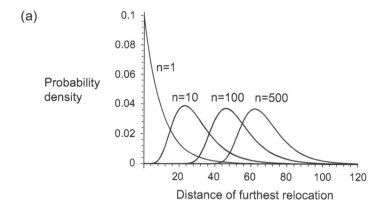

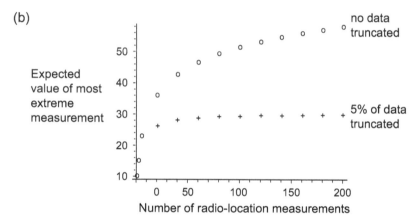

FIGURE 10.1. (a) Solution to equation (10.4) for $\beta = 0.1$, and $N = 1, 10, 100,$ and 500. (b) Expected value of most extreme relocation versus number of relocations. The top curve is given by equation (10.5), and the bottom curve is given by equation (10.8). The value of the localizing tendency parameter β in this example is 0.1.

Taking expectations of the above equation yields the following equation for the expected value for the furthest distance from the den site, $\bar{s}$:

$$\bar{s} = \frac{1}{\beta} \sum_{k=1}^{n_r} \binom{n_r}{k} (-1)^{k+1} \frac{1}{k}, \qquad (10.5)$$

where

$$\binom{n_r}{k} = \frac{n!}{k!(n-k)!}. \qquad (10.6)$$

More details on how to obtain equation (10.5) from equation (10.4) can be found appendix H.

As can be seen in figure 10.1a, the expected value for the furthest measurement $\bar{s}$ grows with sample size, and does not level off as n_r increases. In other words, as the number of relocations increases, the probability of observing a rare, long-range movement continually increases.

We now consider the effect of trimming a certain percentage of the most extreme relocations. The probability that the $n_r - m$th furthest distance measured falls within distance s of the den site is given by

$$P_{n_r}(s) = \sum_{j=0}^{m} \binom{n_r}{j} (1 - \exp(-\beta x))^{n_r-j} \exp(-\beta js). \qquad (10.7)$$

Here the binomial coefficients are "n_r choose j," the number of ways the n_r relocations can be distributed across two groups, one of size j and the other of size $n_r - j$. The first term in the sum corresponds to the situation where all the relocations fall within distance s; the second term in the sum corresponds to the situation where all but one of the measured relocations falls within s, and so on.

As we show in appendix H, the expected value of the distance within which the $n_r - m$th furthest distance measured falls is

$$\bar{s}_{tr} = \bar{s} - \frac{1}{\beta} \sum_{j=1}^{m} \sum_{k=0}^{n_r-j} \binom{n_r}{j} \binom{n_r - j}{k} (-1)^k \frac{1}{k+j}, \qquad (10.8)$$

where the subscript tr indicates that a proportion of points have been excluded prior to calculating the MCP. As figure 10.1b shows, when m is chosen to be a fixed percentage of n_r, an asymptotically constant value for $\bar{s}_{tr}$ is obtained.

Thus our analysis demonstrates that even when an individual's underlying movement behavior gives rise to a well-defined home range for the individual, repeated samples of the individual's position results in a continual increase in the distance to the outermost relocations. However, truncating a percentage of the most extreme relocations results in a saturating estimate for the outermost relocation distances.

The implication of the above results for MCP-based estimates of range size is that conventional MCPs will tend to continually grow in size as the number of relocations increases, making it difficult to characterize the extent of an individual's home range. However, while conventional MCP estimates are problematic, trimmed MCPs can provide a reliable estimate of home range size. The heuristic explanation for this result is simply that as the number of relocations increases, the number of trimmed relocations will also increase, resulting in a more stable characterization of the individual's home range.

For example, figure 10.2 shows MCPs calculated from either 100 relocations (panels a and c) or 500 relocations (panels b and d), for an individual whose

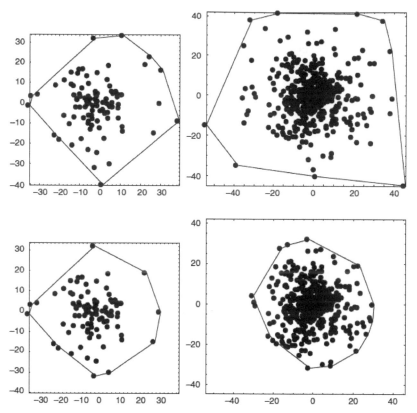

FIGURE 10.2. Minimum convex polygon (MCP) method for relocation data. Note that the figures are sized to give comparable scales between frames. Points show relocation data generated from the two-dimensional Holgate-Okubo localizing tendency model (eq. (3.12)), with $\beta = 0.1$. (a), (b) Conventional MCPs encompassing all relocations when the number of relocations (n_r) is (a) 100, and (b) 500. The home range areas calculated from the MCPs are 32 km^2 and 67 km^2, respectively. (c), (d) 5% trimmed MCPs encompassing 95% of relocations for (c) $n_r = 100$ and (d) $n_r = 500$. Unlike the conventional MCP estimates that differ by a factor of 2, the areas of the trimmed MCPs are very similar, 26 km^2 and 27 km^2, respectively.

underlying movement behavior is governed by the localizing tendency home range model. In panels a and b, conventional MCPs have been calculated. As can be seen in the figure, the estimate of home range size obtained from a conventional MCP increases significantly when the number of relocations of the individual increases, almost doubling from 32 km^2 ($n_r = 100$) to 62 km^2 ($n_r = 500$). Panels c and d show trimmed MCPs for the two datasets in which 5% of the outermost points were excluded. In contrast to the conventional MCP estimates, the trimmed MCP estimates of home range size are similar, 26 km^2 and 27 km^2 respectively.

Upper Bound for the MCP

Equations (10.5) and (10.8) show that the expected distance to the $m/n_r\%$ most extreme relocations in the one-dimensional case is inversely proportional to the bias β in the mechanistic model (eq. (3.9))—the lower the bias the more widespread the relocation observations.

Table 10.1 shows the relationship between the expected extreme relocation distance and inverse bias for a variety of different truncating proportions $100m/n_r\%$. For the more biologically relevant two-dimensional case, it is possible to use equation (3.14) to calculate the distance s within which $100(1 - m/n_r)\%$ observations are expected to fall. In turn, an upper bound for the approximate $100(1 - m/n_r)\%$ MCP area can be found as the area of a circle with radius s. Specifically, $100(1 - m/n_r)\%$ of observations should fall within distance $s = r/\beta$, where

$$2 \int_0^r E_1(\rho)\rho \, d\rho = 1 - m/n_r \qquad (10.9)$$

and $E_1(\rho)$ is the exponential integral (see eq. (3.15)). Solving this for r gives the upper bound for the approximate area of the $100(1 - m/n_r)\%$ MCP as $\pi r/\beta^2$. We expect this upper bound to give a better approximation when the total number of relocations n_r is large, and thus the MCP region is more likely to be approximately circular. Note that the upper bound on the approximate MCP area almost triples when the trimming percentage is reduced from 10% to 2%.

We tested the sensitivity of the predictions from table 10.1 to the number of relocations (n_r) used to calculate the 5% trimmed MCP by generating ten relocation datasets containing either 100, 200, or 500 relocations, and then calculating a corresponding 5% trimmed MCP for each dataset. The results, shown in figure 10.3, indicate that the MCP areas appear to asymptotically approach the bound (eq. (10.9)), but only for very large n_r.

TABLE 10.1. The expected maximum distance observed from the den site and upper bound for the approximate MCP area depending on the percentage of outliers trimmed from the relocation dataset. See also figure 10.3.

Percent trimmed	Expected maximum distance $\bar{s}_{tr}$	Approximate upper bound for MCP area
2	$3.83/\beta$	$62.3/\beta^2$
5	$2.97/\beta$	$38.7/\beta^2$
10	$2.28/\beta$	$24.4/\beta^2$

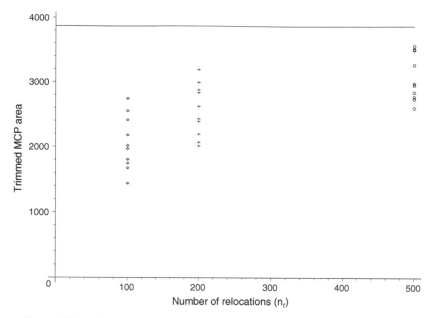

FIGURE 10.3. Relationship between 5% trimmed MCP area and the number of relocations (n_r), given as 100, 200, and 500. The data were generated with a radius taken
from the two-dimensional Holgate-Okubo model (eq. (3.12)), with $\beta = 0.1$ and a
randomly chosen angle. Ten datasets were generated for each of the three sizes. For
an example of the 95% MCP calculations, see figure 10.2. The straight line shows
the upper bound for the approximate MCP area of $38.7/\beta^2 = 3870$ square units (see
table 10.1).

Our analysis thus far has focused on the statistical properties of the minimum
convex polygon method. We now turn our attention to two alternate metrics of
space use: mean-squared and mean-absolute displacement.

10.2. MEAN-ABSOLUTE AND MEAN-SQUARED
DISPLACEMENT

In contrast to MCP-based estimates of home range size that depend on the spatial distribution of outermost relocations, the mean-absolute and mean-squared
displacements relate to the average pattern of space use by an individual. These
statistics feature extensively in the mathematical analysis of correlated random
walks (Kareiva and Shigesada 1983; Othmer et al. 1988; Turchin 1991; Holmes
1993; Turchin 1998). Interestingly, these statistics were also used in early analyses of animal home ranges (e.g., Calhoun and Casby 1958), but have largely
been abandoned since the introduction of the MCP.

Our main reason for examining these alternative metrics of space use is that they can provide insight into the underlying movement behavior of an individual. For example, as discussed in chapter 2, if an animal moves randomly with no autocorrelation in movement direction, then its pattern of space use is given by a spreading Gaussian distribution centered on the initial location of the individual (see figure 2.9). The mean-squared displacement associated with form of movement is simply variance of this distribution. From equation (2.27) we see that this is $4dt$. In other words, if an animal moves randomly in two dimensions its mean-squared displacement will increase linearly over time at $4dt$. A similar analysis indicates that the mean-squared displacement of an individual moving randomly in one dimension increases at rate $2dt$.

As in section 10.1, we investigate the mean-absolute and mean-squared displacements for an individual whose movements are governed by the localizing tendency home range model. For simplicity, we first consider the mean-squared displacement for movement in a single space dimension x. Given a one-dimensional probability density function $u(x, t)$ for the location of the individual as a function of space and time, the mean-absolute (M_1) and mean-squared (M_2) displacements are defined as

$$M_1 = \int_{-\infty}^{\infty} |x| u(x, t) \, dx, \tag{10.10}$$

$$M_2 = \int_{-\infty}^{\infty} x^2 u(x, t) \, dx. \tag{10.11}$$

Multiplying the space use equation for the localizing tendency model (eq. (10.1)) by x^2 and integrating yields

$$\int_{-\infty}^{\infty} x^2 \frac{\partial u}{\partial t} \, dx = d \int_{-\infty}^{\infty} x^2 \frac{\partial^2 u}{\partial x^2} \, dx + c \int_{-\infty}^{\infty} \text{sgn}(x) x^2 \frac{\partial u}{\partial x} \, dx. \tag{10.12}$$

Exchanging integration with differentiation on the left-hand side and integrating by parts twice on the right-hand side, we obtain the following equation for the mean-squared displacement:

$$\frac{dM_2}{dt} = 2d - 2cM_1. \tag{10.13}$$

A similar approach gives an equation for the mean-absolute displacement as

$$\frac{dM_1}{dt} = 2du(0, t) - c. \tag{10.14}$$

If the individual is initially located at its home range center, then both the mean-absolute and mean-squared displacements of the individual will initially be zero. In this situation, the initial conditions for equations (10.13) and (10.14) are $M_2(0) = M_1(0) = 0$. Since $M_1(t)$ is initially close to zero, M_2 initially increases at rate $2d$, the same as that of a randomly moving individual (see above). However, as can be seen from equation (10.13), the presence of a localizing tendency ($c > 0$) causes the rate of growth in M_2 to decline over time as M_1 increases. Eventually M_2 reaches steady-state when the individual's mean-absolute displacement, M_1, reaches $d/c = 1/\beta$. This can be verified by substituting equation (10.2) into equation (10.10). Substituting equation (10.2) into equation (10.11) yields the steady-state value for the mean-squared displacement $M_2 = 2d^2/c^2 = 2/\beta^2$.

Figure 10.4 shows numerical simulations of an individual whose movements are governed by the localizing tendency home range model. As predicted by equations (10.14) and (10.13), the mean-absolute and mean-squared displacements both converge to their asymptotic values—$1/\beta$ and $2/\beta^{-2}$, respectively.

Thus in contrast to a randomly moving individual, the mean-absolute and mean-squared displacements for an individual exhibiting a localizing tendency do not continually increase with time, but instead saturate at values that reflect the magnitude of the localizing tendency β. One implication of this result is that estimates of an individual's mean-squared or mean-absolute displacement provide alternative, statistically robust estimates of home range size that, like trimmed MCPs, do not continually grow with the number of relocations.

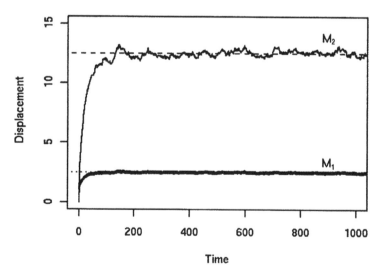

FIGURE 10.4. Numerical simulations of the Holgate-Okubo localizing tendency model (eq. (10.1)) show initial growth in the mean-absolute (M_1) and mean-squared (M_2) displacements followed by a leveling off to a constant value. As predicted by equations (10.13) and (10.14), The slope for $M_2(t)$ is initially $2d$, the value of $M_2(t)$ asymptotes to $2\beta^{-2}$, and the value of $M_1(t)$ asymptotes to β^{-1}.

A further implication is that, since the mean-absolute and mean-squared displacements are simple functions of the localizing tendency parameter β of the Holgate-Okubo model, these measures are sufficient statistics for estimating the strength of an individual's localizing tendency. However, in order to make use of this result, we must generalize the above result to movement in two dimensions.

Displacement in Two Dimensions

In two dimensions, the mean-squared and mean-absolute displacements are defined as

$$M_2 = \int_{-\infty}^{\infty} \int_{-\infty}^{\infty} (x^2 + y^2) u(x, y, t) \, dx dy, \tag{10.15}$$

$$M_1 = \int_{-\infty}^{\infty} \int_{-\infty}^{\infty} \sqrt{x^2 + y^2} u(x, y, t) \, dx dy. \tag{10.16}$$

where $u(x, y, t)$ is the two-dimensional probability density function for the location of the individual. In cases where the pattern of space use is radially symmetric such as the localizing tendency home range model (see eq. (3.12)), the mean-squared and mean-absolute displacements can also be expressed in radial coordinates:

$$M_2 = \int_0^{2\pi} \int_0^{\infty} r^2 u(r, t) r \, dr \, d\theta, \tag{10.17}$$

$$M_1 = \int_0^{2\pi} \int_0^{\infty} r u(r, t) r \, dr \, d\theta, \tag{10.18}$$

where θ is the angle and r the distance from the home range center.

Following a similar procedure to the one-dimensional analysis, substituting the space use equation (eq. (3.12)) into equation (10.18) yields the following equation for the mean-squared displacement:

$$\frac{dM_2}{dt} = 4d - 3cM_1, \tag{10.19}$$

with initial condition $M_2(0) = M_1(0) = 0$. It is also possible to obtain a similar expression for the mean-absolute displacement; however, the equation is considerably more complex than equation (10.19), so we have omitted it here. Note that as in the one-dimensional case, M_2 initially increases at rate $4dt$, the same rate as that of a randomly moving individual.

Substituting the two-dimensional steady-state pattern of space use (eq. (3.12)) into equations (10.17) and (10.18) allows us to calculate the steady-state mean-absolute displacement and mean-squared displacement. At equilibrium,

$M_1 = 4d/(3c)$ and $M_2 = 3d^2/c^2$, respectively. Thus as in the one-dimensional case, the two-dimensional steady-state mean-absolute and mean-squared displacements are simple functions of the individual's localizing tendency, β.

As we have shown, mathematical analysis of mechanistic home range models can provide insight into the relationship between distributions of displacement distances and the underlying movement behavior of individuals. For simplicity, we analyzed the statistics of displacement distances arising from the Holgate-Okubo localizing tendency model. As we saw in chapter 3, the ability of this model to characterize observed home range patterns is relatively poor (see figure 3.10). Nonetheless, the ability to quickly and easily estimate the magnitude of the individual's localizing tendency β from relocation data can serve as a useful prelude to formulating more complex home range models such as those used in chapters 4 and 7. For example, if the estimate of β obtained from the distribution of mean-squared displacement distances is higher than that calculated from the distribution of mean-absolute displacement distances, then the individual's home range is more compact than expected from the localizing tendency model, implying that the individual's localizing tendency increases as its distance from its home range center increases.

10.3. SUMMARY

In this chapter, we analyze the distribution of displacement distances predicted by the Holgate-Okubo localizing tendency model described in chapter 3. Analysis of the distribution of extreme displacements arising from this model shows that truncating a proportion of the furthest relocations before fitting a minimum convex polygon to a dataset yields estimates of home range size that are more robust to variation in the number of relocations obtained for the individual. Analysis of the model's predictions for the mean-absolute and mean-squared displacements of an individual illustrate how these statistics yield insight into an individual's underlying movement behavior. In contrast to those for a randomly moving individual, the mean-absolute and mean-squared displacements arising from the Holgate-Okubo localizing tendency movement model do not continually increase with time, but instead saturate at values that depend on the magnitude of the localizing tendency. These relationships can be used to translate measurements of an individual's mean-absolute or mean-squared displacement into a corresponding estimate for the magnitude of the individual's localizing tendency.

CHAPTER ELEVEN

ESS Analysis of Movement Strategies

Analyzing the Functional Significance

of Home Range Patterns

In essence, the preceding chapters of this book have focused on the following two questions: Given a set of movement rules for individuals, what pattern of home ranges should we expect to see on a particular landscape? And how well do these predicted patterns match observed patterns of space use and scent marking in carnivore populations? As we have seen, the answer to these questions is far from trivial, requiring the use of a formal mathematical approach to translate underlying stochastic movement processes into resulting patterns of space use. As we showed in chapters 2–8, even with fixed rules of movement the pattern of space use resulting from a given movement strategy will vary, depending on proximity and spatial arrangement of neighbors and the underlying spatial heterogeneities that influence movement such as landscape terrain and the spatial distribution of resources; and these kinds of influences can account for aspects of intra- and interspecific variation in home range patterns.

In this chapter, we turn our attention to the related question, why do individuals have the movement strategies they do? As with other aspects of an animal's behavior, the classic approach to answering this question has been to develop models that examine the costs and benefits associated with different patterns of space use (Kodric-Brown and Brown 1978; MacLean and Seastedt 1979; Hixon 1980; Myers et al. 1981; reviewed by Schoener (1983) and Adams (2001)). However, the representations of space use in these models has generally been implicit, represented by the statement "an individual occupies an area of size x" (though see Stamps et al. (1987) and Adler and Gordon (2003)). As a result, no distinction is made between an individual's movement strategy and the pattern of space use that results from the movement strategy being played out on a particular landscape with a given set of neighbors following a similar movement strategy.

The distinction between the underlying rules governing the movement behavior of individuals and the pattern of space use that results from them acquired through the use of mechanistic home range models is important for two reasons. First, as noted above, the relationship between the two is not straightforward and, as a result, certain aspects of observed inter- and intra specific variation in home range patterns can be explained by external influences such as the proximity of neighbors and various forms of spatial heterogeneity, rather than by any difference in the underlying behavioral rules of movement *per se*. Second, models that scale mechanistically from underlying rules of movement to the resulting pattern of space use can incorporate the important biological realities that both an individual's interactions with other individuals and the information it has about its surrounding environment are spatially localized. In this chapter, we use the mechanistic home range model framework to examine the functional significance of different movement behaviors, and determine evolutionarily stable movement strategies that yield patterns of space use that are immune to invasion by groups with alternate behaviors.

11.1. EVOLUTIONARILY STABLE MOVEMENT STRATEGY FOR INTERACTING WOLF PACKS

The biological context for the analysis we present here is the patterns of space use observed in wolf and deer populations in northeastern Minnesota described earlier in chapter 9. Following the approach of Lewis and Moorcroft (2001), we analyze the benefits and costs of the different movement strategies that underpin the mechanistic home range models explored in previous chapters, considering their consequences for a trade-off between utilization of an underlying prey resource and avoidance of hostile neighbors. For simplicity, as in chapter 6, we consider a pairwise interaction between two equal-sized packs, U and V, in a single space dimension x, whose home range centers are located at the edges of a [0, 1] domain and whose patterns of space use are described by the functions $u(x)$ and $v(x)$ respectively (figure 11.1). Although clearly idealized, this scenario retains most of the elements of the more biologically realistic case of multiple packs interacting in two space dimensions.

Fitness Functions

Suppose that the costs and benefits associated with individuals adopting a particular movement strategy are determined by the effects of the resulting pattern of space use on resource acquisition and on the frequency of aggressive interactions with the neighboring pack. While intra-pack interactions among wolves do occur, as in many carnivore societies, wolf packs have strong dominance hierarchies, with alpha females and males dominating the behavior of subordinates

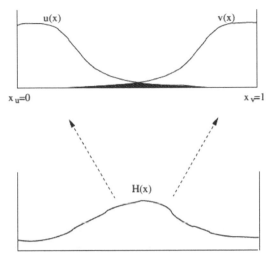

FIGURE 11.1. Schematic of the scenario used to analyze the costs and benefits of different movement strategies. Two equal-sized packs (U and V) move and interact on a one-dimensional [0, 1] domain competing for an underlying prey resource H(x), whose dynamics are given by equations (9.1)–(9.5). The fitness of individuals in each pack varies as a function of the pack's rate of prey intake and the frequency of aggressive encounters, which varies as a function of the degree of home range overlap between the two packs (the shaded area in the figure); see equation (11.1).

(Sheldon 1992). Although subordinates sometimes attempt to mate, this is rarely successful and females typically retain the alpha status for several years (Mech 1966; Peterson et al. 1984; Ballard et al. 1987; Fuller 1989). We therefore propose that a reasonable initial assumption is that each pack operates as a cohesive unit, maximizing the expected number of offspring produced in a single year by the alpha female.

Accordingly, as our measure of fitness, we use the expected number of offspring produced in a single year by a pack's alpha female, that is, the reproductive ratio (geometric growth rate) of the pack R_u:

$$R_u = [\text{survivorship}] \cdot [\text{offspring produced}], \qquad (11.1)$$

where survivorship is the probability that the alpha female survives the year to breed in spring, and the number of offspring produced is the number of pups that survive to weaning, given that the alpha female breeds.

Suppose that yearly offspring production is proportional to prey intake:

$$[\text{offspring produced}] = \sigma\psi \int_0^1 u(x)H(x)dx, \qquad (11.2)$$

where $u(x)$ is the pack's pattern of space use, σ is rate of conversion of prey into offspring, ψ is the prey encounter rate between wolves and deer, and $H(x)$ is the average prey density during the year for packs with patterns of space use $u(x)$ and $v(x)$ (see figure 11.1). $H(x)$ is calculated using the simple model for the spatial distribution of white-tailed deer, used in chapter 9 (eqs. (9.1)–(9.5)).

In addition, suppose that the probability of a wolf being killed as a result of inter-pack aggression is proportional to the local encounter rate between individuals in the two packs $\delta u(x)v(x)$. The overall mortality rate is then given by the integral of this quantity over the whole domain:

$$[\text{survivorship}] = \exp\left(-\delta \int_0^1 u(x)v(x)\,dx\right). \tag{11.3}$$

Substituting equation (11.2) and (11.3) into equation (11.1) yields the following expression for basic reproductive ratio R_u of the alpha female of the U pack:

$$R_u = \underbrace{\exp\left(-\delta \int_0^1 u(x)v(x)\,dx\right)}_{\text{survivorship}} \cdot \underbrace{\sigma\psi\left(\int_0^1 u(x)H(u(x),v(x))\,dx\right)}_{\text{offspring produced}}. \tag{11.4}$$

For ease of analysis, we consider the logarithm of this quantity, that is, the reproductive rate r_u:

$$r_u = \log(\sigma\psi) - \delta \int_0^1 u(x)v(x)\,dx + \log\left(\int_0^1 u(x)H(u(x),v(x))\,dx\right). \tag{11.5}$$

The reproductive rate of pack V, r_v, is given by interchanging $u(x)$ and $v(x)$ in the above formula. We refer to the reproductive rates r_u and r_v as "fitness functions."[1]

Thus the costs and benefits that accrue from individuals adopting a given movement strategy depend on the dynamics of the deer population given the patterns of space use by the two groups, the relationship between resource intake and offspring production, and the costs of aggressive interactions between individuals in neighboring packs. Parameters for the model of deer spatial distribution were chosen to give realistic estimates for the deer recruitment rate and mortality due to predation (see appendix I). The prey encounter rate and conversion of prey into offspring (σ and ψ respectively) form an additive

---

[1] A more complete measure of fitness would also account for the degree of relatedness between the offspring and the alpha female, the future reproductive potential of the alpha female due to reproduction in later years, and variation in population size. However, since these introduce additional complications to the model, we do not pursue them further here.

constant $(\log(\sigma \psi))$ in equation(11.5), hence their values do not affect the relative fitness of different movement strategies; however, we can estimate $\sigma \psi$ by calculating reasonable values for the fitness function (eq. (11.5)). Suppose the deer are at carrying capacity $H = 1$ and there are no hostile interactions with neighbors (mortality parameter $\delta = 0$). Under these conditions we assume that the basic reproductive ratio is approximately 5, i.e., a healthy alpha female has about four pups that survive to weaning (Van Ballenberghe et al. 1975; Fuller 1989). Using equation (11.4), we observe that the basic reproductive ratio is given by $R_u = \sigma \psi = 5$. This is the value that we use when calculating the fitness surfaces in the remainder of the chapter.

The value of the mortality parameter δ is difficult to estimate directly, since the overall mortality rate will depend on the level of home range overlap. However, a recent estimate suggests that the cost of inter-pack aggression is high, accounting for 10% of adult mortalities (Mech 1994). We therefore assume that if the two packs interacted uniformly over the region with no avoidance behavior, then each alpha female would have a 50% chance of survival (i.e., $\delta = 0.69$ in eq. (11.3)).

11.2. ANALYSIS

We assume that the parameters describing the wolf-deer interaction, wolf-wolf mortality, and conversion efficiency from deer into offspring are fixed, and determine the movement strategy values that maximize fitness of individuals in the U and V packs assuming a symmetric competitive game between the two packs. Increasing space use by a pack involves a trade-off between greater resource acquisition that comes from increasing space use, and avoiding encounters with neighbors by reducing space use. We begin by determining the evolutionarily stable values for the parameters of the Holgate-Okubo localizing tendency model analyzed in chapter 3.

"Pure Home Range" Movement Strategy

In the Holgate-Okubo localizing tendency model, individuals exhibit a fixed magnitude bias in their direction of movement toward their home range center. Since the movement of individuals is unaffected by conspecifics, we refer to this as the "pure home range" case. As we showed in chapter 3, the steady-state pattern of space use that results from this movement strategy is given by the solution of an equation of the following form:

$$\underbrace{\nabla^2 u}_{\text{random motion}} - \underbrace{\beta_u \nabla \cdot [u \vec{\mathbf{x}}_u]}_{\text{directed motion}} = 0, \qquad (11.6)$$

where

$$\int_{\Omega} u(x)dx = 1, \tag{11.7}$$

and where the parameter β reflects the strength of directed motion relative to non-directed motion. As we showed in section 3.4 steady-state solution of the above equations in a single space dimension results in patterns of space use that decline exponentially with distance from the home range center. For the situation explored here, in which two interacting packs U and V have home range centers located at the edges of a $[0, 1]$ domain (see figure 11.1), this yields the following equations for space use:

$$u(x) = \beta_u \exp(-\beta_u x)/(1 - \exp(-\beta_u)) \tag{11.8}$$

$$v(x) = \beta_v \exp(-\beta_v(1 - x))/(1 - \exp(-\beta_v)) \tag{11.9}$$

where β_u and β_v reflect the relative strength of directed motion to non-directed motion of the individuals in the two packs.

The evolutionarily stable pure home range strategy is defined by values of β_u and β_v, such that a change in movement behavior (i.e., a change in either β_u or β_v) will result in reduced fitness for the individuals within that pack. In this sense, these values of β_u or β_v are uninvadable by other values and therefore represent an evolutionarily stable strategy or ESS (Maynard-Smith 1974) of movement.[2]

We calculate the evolutionarily stable movement strategy by numerically evaluating the fitness function (11.5) subject to equations (11.8)–(11.9) for different values of the movement parameters β_u and β_v. For each value of β_v, we determine the value of β_u that gives maximum value of the fitness function for pack U using equation (11.5), shown graphically in figure 11.2a as a line of dots. As can be seen from the figure, the resulting ridge of dots crosses the 1:1 ($\beta_v = \beta_u$) line at a single point (A). Because the packs are symmetric, there is an equivalent ridge describing the best movement behavior for individuals in pack V in response to pack U individuals having movement strategy β_u that crosses the 1:1 line at the same point, though for the purpose of clarity we have omitted this from the figure. At this intersection point, individuals in both packs will do worse if they adopt an alternative behavior, and therefore the coordinates of the intersection point (A) define the values of β_u and β_v that constitute an

---

[2]Note that in technical terms, our ESS criterion is for a "game against the field" as opposed to a pairwise game in which packs attempt to maximize a fitness differential ($r_u - r_v$). Since in natural populations, packs may have up to six neighboring packs (Peters and Mech 1975; Peterson et al. 1984; Ballard et al. 1987; Fuller 1989), we propose that a "game against the field" is the more appropriate caricature of the natural system where a game is being waged in two dimensions against multiple neighbors.

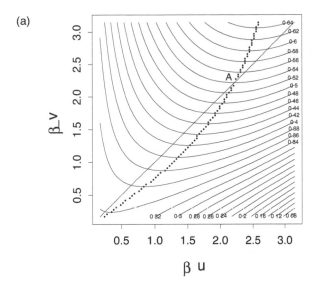

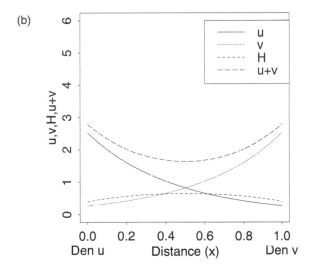

FIGURE 11.2. Evolutionarily stable strategy for the pure home range model $\beta_u = \beta_v = \beta^* = 2.26$ in equations (11.8)–(11.9). (a) Contour lines show the fitness (as given by reproductive rate r_u) of the U-pack individuals (11.5) as a function of the movement strategy of the V pack individuals β_v and the movement strategy for U pack individuals β_u. For every value of β_{sv}, (·) indicates the β_u value yielding the maximum value of r_u. The point at which the ridge ($\cdots$) crosses the 1:1 line (point A) defines the evolutionarily stable movement strategy (ESS). The fitness of both packs (r_u and r_v) at the ESS is 0.55. The pattern of space use resulting from the evolutionarily stable movement strategy is shown in panel (b). (b) Patterns of expected space use $u(x)$ and $v(x)$ and the resulting distribution of deer $H(x)$ calculated using equation (9.5) with $\lambda = 2$ and $\psi = 0.15$. Notice that the deer distribution $H(x)$ is highest where the total density of space use is lowest—at the boundary between the two groups.

evolutionarily stable movement strategy ($\beta_u^* = \beta_v^* = 2.26$). The patterns of space use associated with the ESS solution of the pure home range model are shown in figure 11.2b. Since space use by each pack declines exponentially as a function of distance from its home range center, the combined space use by the two packs is considerably lower at the boundary between the two packs (figure 11.2b). This lowers the predation rate in this region and, as a result, deer densities are higher here, forming a prey "reservoir" at the boundary between the two packs.

"Pure Territorial" Movement Strategy

We now consider an alternative movement strategy in which individuals exhibit an avoidance response to foreign scent marks. Specifically, we consider the model analyzed in section 6.2 of chapter 6, in which individuals scent-mark as they move and exhibit an avoidance of foreign scent marks but do not engage in overmarking (the parameter m set to zero in eqs.(4.17)–(4.18)). The steady-state patterns of space use that result from this movement strategy are given by the solution of the following non-dimensionalized equations:

$$\nabla^2 u - \beta_{su} \nabla \cdot \left[u \vec{\mathbf{x}}_u q \right] = 0, \tag{11.10}$$

$$\nabla^2 v - \beta_{sv} \nabla \cdot \left[v \vec{\mathbf{x}}_v p \right] = 0, \tag{11.11}$$

$$u - p = 0, \tag{11.12}$$

$$v - q = 0, \tag{11.13}$$

where parameters β_{su} and β_{sv} reflect the strength of directed motion caused by encounters with foreign scent marks relative to non-directed motion of the individuals in the two packs. As we determined in section 6.2, when implemented on a one-dimensional $[0, 1]$ domain, this yields the following equations for the steady-state pattern of space use:

$$\frac{\partial u}{\partial x} = -\beta_{su} uv, \tag{11.14}$$

$$\frac{\partial v}{\partial x} = \beta_{sv} uv. \tag{11.15}$$

We refer to this as the "pure territorial" case, since spatial partitioning only occurs as a result of encounters via foreign scent marks. If the interaction is symmetric, both packs have the same movement strategy (i.e., $\beta_{su} = \beta_{sv} = \beta_s$), and the pattern of space use by the two packs is given by two logistic equations:

$$\frac{\partial u}{\partial x} = -\beta_s u \left(2 - u \right), \quad \frac{\partial v}{\partial x} = \beta_s v \left(2 - v \right) \tag{11.16}$$

(see chapter 6).

We calculate the ESS for the pure territorial model using the same procedure that we use to determine the ESS for the pure home range movement strategy, that is, by determining the value of parameters β_{su} and β_{sv} such that any change in their value by the respective pack results in reduced fitness for that pack (figure 11.3). As can been seen from the figure, the symmetric nature of the interaction between the two groups means that there is a single evolutionarily stable territorial movement strategy in which individuals in both packs exhibit the same degree of scent mark avoidance (i.e., $\beta_{su} = \beta_{sv} = \beta_s^* = 2.95$).

The patterns of space use and resource density associated with this movement strategy are shown in figure 11.3b. Comparison of the evolutionarily stable patterns of space use for the two cases shows that a pure territorial movement strategy (Case 2) results in higher fitness for both packs than a pure home range movement strategy (Case 1) ($r_u, r_v = 0.62$ versus $r_u, r_v = 0.55$; see figures 11.2 and 11.3). This is due to the logistic pattern of space use by the two packs that arises from a pure territorial movement strategy, which both increases the benefits and reduces the costs experienced by each pack. The increased benefits arise because the logistic pattern of space use by each pack results in a uniform pattern of total space use and predation intensity across the domain, leading to a more complete utilization of the available prey resource than in the pure home range model (compare figures 11.2b and 11.3b). The reduction in costs arises because the home ranges of the two packs now have sharper edges, which reduces home range overlap, lowering the encounter rate between the two packs, increasing survivorship (see figures 11.2b and 11.3b).

Mixed Territorial/Home Range Strategies

We now consider whether these pure strategies can be invaded by a more general strategy that combines a localizing tendency with an avoidance response to foreign scent marks. We do this by considering a mixed territorial/home range movement strategy in which individuals can exhibit both a localizing tendency and an avoidance response to foreign scent marks. The equations for space use for this mixed strategy model are:

$$\nabla^2 u - \nabla \cdot \left[u \vec{\mathbf{x}}_u (\beta_u + \beta_{su} q) \right] = 0, \tag{11.17}$$

$$\nabla^2 v - \nabla \cdot \left[v \vec{\mathbf{x}}_v (\beta_v + \beta_{sv} p) \right] = 0, \tag{11.18}$$

$$u - p = 0, \tag{11.19}$$

$$v - q = 0. \tag{11.20}$$

The pure home range (Case 1) and territorial movement (Case 2) strategies are special cases of the above model, obtained by setting either β_{su} and β_{sv} to zero or β_u and β_v to zero, respectively.

(a)

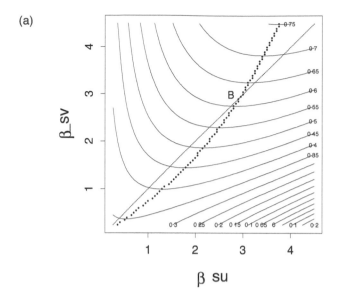

(b)

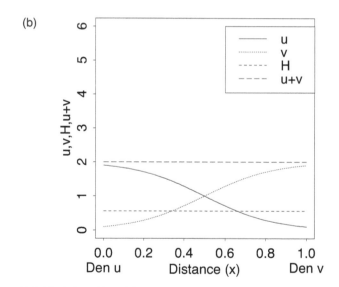

FIGURE 11.3. Evolutionarily stable strategy for pure territorial model (eqs. (11.14)–(11.15)). (a) Contour lines show r_u, the fitness function for individuals in pack U (eq. ((11.5)) as a function of the movement strategy of pack V individuals (β_{sv}) and pack U individuals (β_{su}). For every value of β_{sv}, ($\cdot$) is the β_{su} value yielding the maximum value of r_u. As in figure 11.2, the point at which the ridge ($\cdots$) crosses the 1:1 line (point B) defines the evolutionarily stable pattern of space use by the two packs ($\beta_{su} = \beta_{sv} = \beta_s^* = 2.95$), which yields a reproductive rate of $r_u = 0.62$. (b) The expected pattern of space use $u(x)$ and $v(x)$ calculated from equation (11.16), and the resulting distribution of deer $H(x)$, calculated by (9.5) with $\lambda = 2$ and $\psi = 0.15$. Notice that the deer distribution $H(x)$ is uniform throughout the domain.

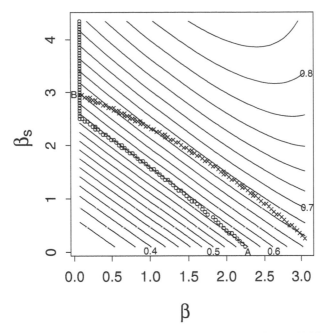

β

FIGURE 11.4. Fitness surface for the mixed strategy model (eqs. (11.17)–(11.20)) when both packs are playing the same strategy. Contour lines indicate the fitness of individuals belonging to either pack along the manifold ($\beta_u = \beta_v = \beta$ and $\beta_{su} = \beta_{sv} = \beta_s$). Points A and B are the pure home range and pure territorial ESSs respectively. The ridge of o's indicates values of β and β_s for which the fitness of individuals within a pack will decrease if the home range component of their movement strategy (β_u or β_v depending on the pack) deviates from β. The ridge of x's shows values of β and β_s for which the fitness of individuals within a pack will decrease if the territorial component of their movement strategy (β_{su} or β_{sv} depending on the pack) deviates from β_s. A necessary condition for an ESS is that the ridges of x's and o's intersect. The two ridges intersect at point B ($\beta = 0, \beta_s = 2.95$), the pure territorial ESS, indicating that the pure territorial ESS (figure 11.3) also is an ESS for the mixed strategy model.

Determining the ESS for the above movement model is more complex, as it involves potential changes in both components of the movement behavior by each pack. A necessary (though not sufficient) criterion for a mixed ESS is that a pack's fitness decreases if it changes its value of β or its value of β_s. Figure 11.4 shows the fitness surface for the mixed strategy model when both packs have the same strategy (i.e., when $\beta_u = \beta_v = \beta$ and $\beta_{su} = \beta_{sv} = \beta_s$). The ridge of o's shows values of β and β_s for which the fitness of a pack will decrease if it deviates from β; the ridge of x's shows values of β and β_s for which a pack's fitness will decrease if it deviates from β_s. A mixed strategy ESS would require an intersection of the two ridges in the positive quadrant (i.e., $\beta, \beta_s > 0$). This does not occur: the ridges of o's and x's intersect at $\beta = 0$,

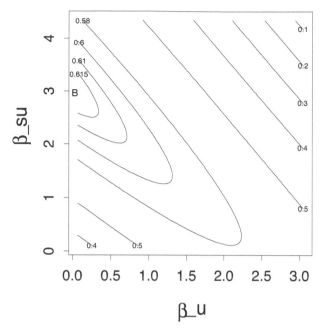

FIGURE 11.5. Fitness payoff r_u for pack U as a function of β_u and β_{su} given that pack V has the pure territorial strategy identified in figure 11.3 ($\beta_v = 0$, $\beta_{sv} = 2.95$). Point B in the figure corresponds to B in figure 11.4, the pure territorial ESS. At point B, the strategy and fitness of pack U matches that of pack V. All other possible strategies by pack U yield lower fitness, demonstrating that pure territorial strategy $\beta = 0$, $\beta_s = 2.95$ is stable to invasion and is thus an ESS for the mixed strategy model, equations (11.17)–(11.20).

$\beta_s = 2.95$, the pure territorial ESS (point B in figure 11.4), implying that this is the ESS for the mixed strategy model. Figure 11.5 confirms this, showing how the fitness payoff for pack U varies as a function of β_u and β_{su} given that pack V has the pure territorial strategy defined by point B ($\beta_v = 0$, $\beta_{sv} = 2.95$). At all points other than point B, pack U has a reduced fitness payoff, confirming that the pure territorial strategy ($\beta = 0$, $\beta_s = 2.95$) is also an ESS for the mixed territorial/home range strategy model.

11.3. ROLES OF AGGRESSION AND SIGNALING

The evolutionarily stable movement strategy for the simplified mechanistic home range models analyzed in this chapter uses foreign scent marks to modulate movement toward the den site ($\beta_s^* = 2.95$, $\beta^* = 0$). Alternate strategies that include a fixed bias in movement direction toward the den site ($\beta > 0$) cannot invade this pure territorial movement strategy. A characteristic feature

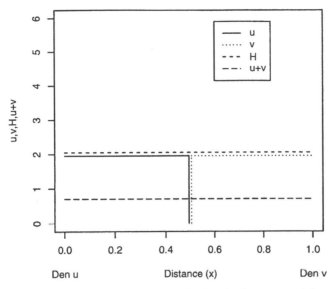

FIGURE 11.6. The optimal movement strategy for the mixed strategy model equations (11.17)–(11.20). Figure shows the pattern of space use that arises when $\beta_{su}, \beta_{sv} \to \infty$, which results in complete and equitable partitioning of the available resource with no overlap between the home ranges of the two packs. This results in higher fitness for individuals in both packs than the evolutionarily stable strategy $\beta_{su}, \beta_{sv} = \beta_s^*$ shown in figure 11.3. However, this optimal strategy is susceptible to invasion by packs that adopt a lower value of β_s.

of the pure territorial ESS is that it gives rise to a logistic pattern of space use by the two packs that results in a spatially uniform pattern of prey utilization and prey density (see figure 11.3).

The optimal, as opposed to evolutionarily stable, movement strategy would be to let β_{su} and β_{sv} go to infinity, so that the pattern of space use by each pack $u(x)$ and $v(x)$ given by the solution of equations (11.10)–(11.13) becomes a step function (figure 11.6). When this occurs, there are no aggressive interactions between the neighboring packs and space is completely and equitably partitioned. However, this optimal strategy is not stable and will be invaded by a strategy with reduced sensitivity to scent marks (i.e., a finite β_s movement strategy). This implies that the home range overlap that arises in the evolutionarily stable territorial movement strategy, resulting in aggressive encounters between the two groups (see figure 11.3b), is the necessary condition that confers competitive stability on the patterns of space use by the two packs.

A heuristic explanation for the pure territorial ESS is that in the absence of foreign scent marks, packs with this movement behavior expand their space use via simple diffusion to fill the room available (see figure 11.3b and eqs. (11.10)–(11.13)). In contrast, alternate "pure home range" or "mixed" strategies that contain a built-in bias toward the den site stop expanding when the directed and random components of motion balance (see figure 11.2b and eqs. (11.6), (11.17)–(11.20)), even in the absence of another group. In this sense, pure home range and mixed strategies do not fully avail themselves of the opportunity to invade space previously occupied by neighbors if this space becomes available.

For simplicity, the territorial movement strategy we have analyzed is the version of the conspecific avoidance model in which individuals mark at a constant rate. An important consequence of this assumption is that the spatial distribution of a pack's scent marks is an accurate reflection of its intensity of space use (see eqs. (4.13)–(4.14) and figure 11.3b), and thus constitute an "honest signal" of space use (*sensu* Johnstone 1997). The spatially localized nature of scent marks is an important physical constraint that helps maintain their reliability as an indicator of space use; however, given the inherent conflict of interest between neighboring groups, we might expect that individuals would attempt to misrepresent their intensity of space use by marking at higher rates in disputed areas.

This leads us naturally to consider the full formulation of the scent-marking avoidance model developed in chapter 4, in which individuals increase their scent-marking rate response to encounters with foreign scent marks. As we showed in chapters 5 and 6, this can yield "bowl-shaped" scent-mark distributions and give rise to a "buffer zone"—an area of low space use between the packs (see color plate 6 and figure 6.3). In these cases, scent marks are no longer an "honest" reflection of space use: scent-mark levels are highest at the edges of the territories, where space use declines. An interesting subsequent analysis would be to investigate whether the "pure territorial strategy" will be invaded by "territorial over-marking strategy" in which individuals exhibit an avoidance response to encounters with foreign scent marks but also increase their marking rate in in order to try and "bluff" neighboring packs into retreating.

11.4. SUMMARY

In this chapter, we use a game-theoretic approach to analyze the costs and benefits of different movement strategies for a pair of interacting wolf packs. By modifying their rules of movement, packs adjust their patterns of space use to maximize their reproductive output. This involves a trade-off between maximizing prey intake and minimizing conflict with neighbors. Evolutionarily stable choices of the behavioral parameters yield territories that are immune to invasion by groups with alternate behaviors. In contrast to earlier spatially implicit models of territoriality, the mechanistic home range models analyzed take explicit account of the relationship between individual rules of movement, resulting patterns of space use, and the subsequent fitness of individuals, and capture the important biological reality that an individual's interactions and information about its environment are both spatially localized. Moreover, as we have already seen, their spatially explicit nature also means that mechanistic home range models can be directly tested against empirical home range data, thereby offering a way to formally integrate empirical measurements of animal space use with functional analyses of home range patterns.

CHAPTER TWELVE

Future Directions and Synthesis

The analyses in the preceding chapters suggest a number of avenues for future research that we now consider before synthesizing the main findings of this book.

Temporal Variability

Throughout the book, our analysis of home range patterns has focused on steady-state patterns of space use given by time-independent solutions of the space use equations, implying either a stable or, equivalently, a temporally averaged environment. However, in many carnivore populations, both the biotic and abiotic components of an animal's environment fluctuate on a variety of time scales. For example, as discussed in chapter 7, seasonal variation in small mammal abundance and seasonal ungulate reproduction and migrations gives rise to large-scale shifts in the spatial distribution and abundance of wolf and coyote prey (Van Ballenberghe et al. 1975; Bekoff and Wells 1980; Bowen 1981; Bekoff and Wells 1981; Gese et al. 1996a; Gese et al. 1996b; Ballard et al. 1998; Crabtree and Sheldon 1999), while on shorter time scales, weather-related changes in snow depth and crust condition affect the ease of movement and rates of killing by wolf and coyote packs (Gese and Grothe 1995; Post 1999; Nelson and Mech 1986; Mech and Boitani 2003). Capturing the influence of such temporal dynamics on the movement behavior of individuals and resulting patterns of space use will require considering time-dependent solutions of the space use equations.

The Adaptive Significance of Home Range Patterns

Eight of the twelve chapters in this book (chapters 3–10) focus on analyzing the mechanistic relationship between different movement rules for individuals and their consequences for patterns of space use in carnivores in the belief that gaining a quantitative and reductionist understanding of observed home range patterns is fundamental to developing a general theory of carnivore-space use. Another key aspect of such a theory is understanding the adaptive or functional significance of different movement strategies exhibited by individuals in different populations. As we saw in chapters 5 and 7, even with fixed movement rules, patterns of space use can vary due to changes in an individual's biotic and abiotic environment, and such influences may account for significant portions of the intra- and interspecific variation in the home range patterns seen in figure 1.1. An important additional source of variation in patterns of space use comes from individuals changing their movement behavior in response to the conditions they experience. The capability for such behavioral learning is well accepted in canids and other groups of carnivores (Nel 1999).

We address the issue of behavioral learning in chapter 11 through a game-theoretic analysis of the localizing tendency and conspecific avoidance home range models. The results of this analysis highlight the importance of avoidance responses to foreign scent marks in reducing home range overlap between neighbors, but also the continuing need for some degree of home range overlap in order to confer stability on the partitioning of space between individuals that have competing fitness interests. Although in its infancy, such game theoretic analysis of movement strategies provides a methodology for understanding the functional significance of the home range patterns found in different carnivore populations. Future studies of this kind will want to move beyond the simple pairwise interaction studied in chapter 11 and consider games involving multiple neighbors, the potential for mixed strategy solutions (Maynard-Smith 1974), more sophisticated movement strategies such as state-dependent and environment-dependent foraging strategies (Grunbaum 1998; Grunbaum 2000), dynamic information (Mangel and Clark 1988; Mangel and Roitberg 1989), more complex scent-marking strategies that involve "dishonest signaling" and "bluffing" (Krebs and Dawkins 1984), and the use of landmarks (Mesterton-Gibbons and Adams 2003).

Carnivore Conservation and Management

Our primary motivation for developing this mechanistic approach to home range analysis was to gain biological insight into the causes and consequences of carnivore home range patterns. But this approach has an important further application. Like many other animals carnivore populations are increasingly threatened by habitat loss, hunting, and other forms of human activity (Woodroffe and Ginsberg 1998; Ginsberg 2001; Woodroffe 2001). In the face of these multiple stresses, carnivore populations around the world are increasingly being managed to maintain numbers (Ginsberg and Macdonald 1990; Gittleman et al. 2001).

Although preliminary, the pack removals and additions and resource manipulations performed in chapters 4, 5, and 7 illustrate how mechanistic home range models can be used to predict the consequences of different management actions. More detailed formulations that incorporate responses to resource availability and landscape characteristics (chapter 7) offer even greater scope for designing conservation and management strategies for carnivore populations. There is also growing recognition of the importance of considering the effects of management actions not just on a single "target" species but in a broader community context. For example, as discussed in chapter 9, the introduction of wolves to Yellowstone has had profound effects on both ungulates and other members of the scavenger community, particularly coyotes (Crabtree and Sheldon 1999; Smith et al. 2003). And in Africa, efforts to conserve African wild dogs are complicated by their competitive interactions with lions and hyenas (Creel and Creel 1996; Creel et al. 2001). As our analysis in chapter 9 illustrates, the mechanistic approach advocated here readily extends to considering interactions with prey and competitors, and thus can be used to design conservation and management plans that balance the interests of competing management objectives or conservation goals.

Other Species

The home range model formulations developed in this book reflect the ecological and behaviorial characteristics of carnivore movement, such as den sites acting as a focal point for the movement of individuals, scent-mark avoidance, and overmarking responses following encounters with foreign scent marks. While we have focused on carnivores, these formulations are also likely applicable to a number of other mammalian species that occupy distinct home ranges and utilize scent marks as indicators of territorial occupation, such as lemurs and tamarins (Jolly 1966; French and Cleveland 1984). More generally, the approach we have taken here for carnivores could also be applied to the study of home range patterns in a variety of other taxa, including primates (Waser 1985; Barrett and Lowen 1988), ungulates (Jarman 1974; Clutton-Brock et al. 1982; Boyce et al. 2004), rodents (Ostfeld 1986; Shenbrot et al. 1999), and lizards (Stamps 1977; Schoener and Schoener 1980; Roughgarden 1995). While differences in the underlying ecology of movement in these different animal groups will most likely result in different space use equations, most of the mathematical methods and techniques used in this book are also suitable for analyzing home range patterns in these other taxanomic groups.

Synthesis

The objective of this book has been to develop and analyze a series of mechanistic and spatially explicit home range models for carnivore home ranges. This has involved taking a reductionist view of patterns of space use, treating them as macroscopic patterns that result from underlying individual-level behavioral movement rules.

The reductionist nature of mechanistic home range models enables them to be parameterized and tested against observations of space use collected at a variety of spatial and temporal scales. Until recently, obtaining observations about both an individual's fine-scale movement behavior and patterns of space use was relatively difficult, requiring separate measurement systems. However, modern global-positioning system relocation collars allow for sophisticated sampling schemes that can collect measurements at a variety of temporal scales. For example, collars can be programmed to collect short bursts of high-frequency relocations at intermittent intervals, yielding detailed information on the fine-scale movement behavior of individuals, while continually relocating individuals at low frequency over a period of several months, yielding information about their long-term, large-scale patterns of space use.

In establishing a connection between observed patterns of space use and underlying individual-level descriptions of movement behavior, mechanistic home range models serve as a vehicle for evaluating competing ecological hypotheses for the determinants of carnivore home range patterns. Each set of movement rules and associated parameter values is, in effect, a hypothesis for the animal's observed pattern of space use. As different behavioral responses are incorporated into the movement rules or as the magnitude of these different components changes, the surfaces describing the patterns of space use by individuals also change accordingly.

Our approach to evaluating different models (hypotheses) for space use is based on the principle of sufficiency. That is, we ask: what set of factors must we incorporate into the model to adequately account for observed patterns of space use? Through our analyses of relocation datasets at Cedar Creek, Hanford, and Yellowstone, we examined the influence of a number of factors affecting space use in carnivores. We began by parameterizing a simple localizing tendency home range model using observations of the fine-scale movement behavior of a red fox at Cedar Creek (chapter 3). Subsequent comparison of this model to observed patterns of space use by coyote packs at Hanford highlighted the limited ability of this simple model to characterize the boundaries between neighboring groups, leading us to explore the role that conspecific avoidance plays in determining home range patterns (chapter 4). The improved fit obtained by incorporating scent-mark avoidance into the home range model provides empirical support for the hypothesis that conspecific avoidance, mediated by avoidance responses to foreign scent marks, plays a significant role in determining observed patterns of coyote patterns of space use. Numerical and analytic investigations into the qualitative properities and predictions of the conspecific avoidance home range model (chapters 5 and 6) yields insight into the patterns of space use and scent marking observed in different carnivore populations, including an ecological explanation for the border and hinterland marking patterns and role of population density in affecting the conditions under which "buffer zones" form between territories. Further analysis showed how the conspecific avoidance model could be extended to gain biological insight into landscape heterogeneities affecting patterns of coyote space use in Yellowstone National Park (chapter 7). Terrain heterogeneity and prey availability were evaluated as alternative hypotheses for the observed spatial distribution of coyote home ranges by formulating extended versions of the conspecific avoidance model that incorporated movement responses to these two different sources of hetergeneity. Comparison of the predictions of these two different sources of heterogeneity. Comparison of the predictions of these models against the observed patterns of coyote relocations in Yellowstone showed that movement responses to spatial heterogeneity in prey availability provided a substantially better explanation for the observed patterns of space use than movement responses to terrain heterogeneity. We then showed that the fitted mechanistic home range model incorporating responses to prey availability correctly predicted the observed shifts in the patterns of coyote space use that occurred in response to natural demographic perturbation.

The initial stage of a mechanistic home range analysis, in which an underlying stochastic movement model is formulated for the movement behavior of individuals (see chapters 3–4 and 7–8), is similar to the individual-based modeling approach advocated by DeAngelis, Gross, and others (DeAngelis and Gross 1992; Conroy et al. 1995; Dunning et al. 1995). The first model of this kind for carnivores was developed in the late 1960s (Siniff and Jessen 1969), and since then, a number of models of this kind have been developed for different carnivore species (Benhamou 1988; Benhamou 1989; Lima and Zollner 1996; Comiskey et al. 1997; Zollner and Lima 1999; Ahearn et al. 2001; MacDonald and Rushton 2003). In contrast to the mechanistic home range model approach we use here, where we formulate equations for the patterns of space use that result from the underlying movement model, in the individual-based simulation approach expected patterns of space use are calculated by performing repeated simulations of the underlying movement model. Mechanistic home range models offer two principal advantages over a pure, simulation-based approach. First, the formulation of the space use equations avoids the computationally intensive procedure of repeated numerical simulations to calculate the patterns of space use that result from the underlying movement model. Second, the mathematical compactness of the space use equations yields analytic insight into the patterns of space use that arise from the underlying movement model. For example, analysis of the conspecific avoidance model equations showed how an overmarking response to foreign scent marks gives rise to buffer zones between territories, and can also generate "bowl-shaped" scent-mark distributions, depending on the magnitude of the overmarking response and the density of home ranges (chapters 5 and 6). In addition, analysis of the Holgate-Okubo localizing tendency model revealed the signature of a localizing tendency that arises in the rate of change in an individual's mean-squared displacement (chapter 10). As these two examples illustrate, the analytic nature of mechanistic home range models yields insight into the "cause-and-effect" relationship between different forms of movement behavior and resulting home range patterns.

The mechanistic and spatially explicit nature of the home range models developed in this book also addresses several limitations of the resource selection approach to empirical home range analysis mentioned in the introduction. First, the mechanistic home range approach avoids the need to define *a priori* "available habitat." This is a critical step in resource selection studies, often involving subjective judgment, that can significantly impact their findings (Johnson 1980; Alldredge & Ratti 1986; Porter and Church 1987; Thomas & Taylor 1990; Aebischer et al. 1993; Arthur et al. 1993; Cooper and Millspaugh 2001; Matthiopoulos 2003). In contrast, mechanistic home range analysis avoids the issue of having to define what areas are available to an individual, in effect, because the underlying model of individual movement behavior determines the likelihood and feasability of an individual moving to a particular location given its current position.

Second, the spatially explicit nature of mechanistic home range models means that they predict actual spatial patterns of space use rather than simply relative rates of habitat utilization obtained from resource selection analyses. This enables mechanistic home range models to account for the constraints that landscape geometry imposes on patterns of space use by individuals, and thus make more complete use of the spatially explicit nature of telemetry datasets. For example, in resource selection analysis, all areas within the region defined as "available" are assumed to be directly accessible to an individual, while on actual landscapes, the patchy spatial distribution of habitats—such as the patches of high small mammal biomass in Lamar Valley seen in figure 7.1— means that individuals frequently traverse less favorable habitat in order to move between favorable areas. In resource selection analysis, the times individuals spend traversing unfavorable areas can register as a degree of selection, rather than as constraint imposed by the spatial geometry of the landscape. In contrast, by explicitly incorporating the process of individual-level movement, mechanistic home range models naturally incorporate the effects of geometric constraints on patterns of space use.

Third, mechanistic home range models are able to incorporate the effects of behavioral factors that also affect patterns of space use by individuals. As seen in chapters 4, 7, and 9, this is particularly important in carnivores where den sites often act as focal points for the movements of individuals and individuals exhibit avoidance responses to conspecifics and competitors. Moreover, as seen in color plates 15 and 16, in addition to influencing current patterns of space use by individuals, the spatial geometry of resources and the constraints imposed by the patterns of space use of neighboring groups can significantly influence the way in which home range patterns shift in response to demographic and environmental perturbations.

An interesting and important future step will be to compare the results of resource selection and mechanistic home range model analyses of identical relocation datasets. A natural metric that can be extracted from mechanistic home range model fits for such a comparison is Turchin's (1998) "Residence Index," the proportion of the probability density function that is associated with each habitat type.

Finally, unlike either resource selection analyses or pure simulation-based approaches, mechanistic home range models can bridge naturally between detailed formulations used to characterize empirical relocation datasets (chapters 3, 4, and 7) and predict observed changes in patterns of space use (chapter 7), and simplified, analytically tractable forms of these same models that yield general insights into patterns of carnivore space use and their adaptive significance (chapters 5, 6, 10, and 11). This ability to move seamlessly between so-called "tactical" and "strategic" (Holling 1966) models of carnivore home ranges promises to yield a theory of carnivore space use that combines reality and precision with generality (Levins 1966; May 1974).

Derivation of the Fokker-Planck Equation
for Space Use

We start with equation (2.9) which describes the change in the probability density function $p(x,t)$ over a time interval of length τ in terms of the redistribution kernel $k(x',x,\tau,t)$

$$p(x,t+\tau) = \int_{-\infty}^{\infty} p(x',t)k(x',x,\tau,t)dx'. \tag{A1}$$

We assume that, in each time step τ, a spatial jump $a = x-x'$ is taken from x' to x, and that direction and length of this jump depends only on the starting location x'. This means the kernel can be rewritten as $k(x',x,\tau,t) = k_a(x',a,\tau,t)$. As shown below, this assumption of the starting point determining the jump leads to the Fokker-Planck (forward Kolmogorov) equation for space use.

To translate the above integrodifferential equation into a differential equation, we expand the terms on the right-hand side of equation (A1) using a Taylor series expansion for $p(x',t)k_a(x',a,\tau,t)$ with respect to the variable x' about x, and change the integration variable from x' to a:

$$
\begin{aligned}
p(x,t+\tau) &= \int_{-\infty}^{\infty} \Big[p(x,t)k_a(x,a,\tau,t) \\
&\quad - a\frac{\partial}{\partial x}\big[p(x,t)k_a(x,a,\tau,t)\big] \\
&\quad + \frac{a^2}{2}\frac{\partial^2}{\partial x^2}\big[p(x,t)k_a(x,a,\tau,t)\big] + \dots \Big] da.
\end{aligned} \tag{A2}
$$

Dividing (A2) by τ and switching the order of differentiation and integration we get

$$
\begin{aligned}
\frac{p(x,t+\tau) - p(x,t)}{\tau} &= -\frac{1}{\tau}\frac{\partial}{\partial x}\int_{-\infty}^{\infty} ap(x,t)k_a(x,a,\tau,t)da \\
&\quad + \frac{1}{2\tau}\frac{\partial^2}{\partial x^2}\int_{-\infty}^{\infty} a^2 p(x,t)k(x,a,\tau,t)da + \dots
\end{aligned} \tag{A3}
$$

since $\int_{-\infty}^{\infty} p(x,t)k_a(x,a,\tau,t)da = p(x,t)$.

We now let τ become small, i.e., we take the limit $\tau \to 0$ so that

$$\frac{\partial p(x,t)}{\partial t} = -\frac{\partial}{\partial x}\left[\lim_{\tau \to 0}\frac{1}{\tau}\int_{-\infty}^{\infty} a k_a(x,a,\tau,t)da\, p(x,t)\right]$$

$$+ \frac{\partial^2}{\partial x^2}\left[\lim_{\tau \to 0}\frac{1}{2\tau}\int_{-\infty}^{\infty} a^2 k_a(x,a,\tau,t)da\, p(x,t)\right] + \ldots \qquad (A4)$$

Defining

$$c(x,t) = \lim_{\tau \to 0}\frac{1}{\tau}\int_{-\infty}^{\infty} a k_a(x,a,\tau,t)da, \quad \text{and}$$

$$d(x,t) = \lim_{\tau \to 0}\frac{1}{2\tau}\int_{-\infty}^{\infty} a^2 k_a(x,a,\tau,t)da, \qquad (A5)$$

we arrive at equation (2.10)

$$\frac{\partial p(x,t)}{\partial t} = -\frac{\partial}{\partial x}\left[c(x,t)p(x,t)\right] + \frac{\partial^2}{\partial x^2}\left[d(x,t)p(x,t)\right]. \qquad (A6)$$

The coefficients $c(x,t)$ and $d(x,t)$ can also be expressed in terms of $k(x',x,\tau,t)$:

$$c(x,t) = \lim_{\tau \to 0}\frac{1}{\tau}\int_{-\infty}^{\infty}(x'-x)k(x',x,\tau,t)dx', \quad \text{and}$$

$$d(x,t) = \lim_{\tau \to 0}\frac{1}{2\tau}\int_{-\infty}^{\infty}(x'-x)^2 k(x',x,\tau,t)dx'. \qquad (A7)$$

Alternative Derivation
of the Space Use Equation

Equation (2.6) can also be derived from a conservation law with specified local *flux* of expected individuals (Murray 1989). While this framework is not related directly to the random walks of individuals, it is a standard approach to formulating advection-diffusion equations. The philosophy underlying this framework is that in the absence of birth and death, the local flux, or "flow" of expected individuals from one point to another, can describe changes in local density. The flux J is a vector with units expected density times velocity. The conservation equation

$$\frac{\partial u}{\partial t} + \frac{\partial J}{\partial x} = 0 \tag{B1}$$

relates the flux to the local density. A derivation of the conservation law from first principles is given in Edelstein-Keshet (1988). The choice of

$$J = J_d + J_c, \tag{B2}$$

where

$$J_d = -\frac{\partial}{\partial x}(d(x,t)u), \quad J_c = c(x,t)u \tag{B3}$$

yields the advection-diffusion equation (2.10). The formulation for J_d indicates that the net effect of the random component of motion is to cause a net flux of individuals down the locally weighted gradient in population density. Here $d > 0$ is the weighting function, with large d indicating rapid flux from high to low density. Recalling that the flux is defined to be velocity times density, we observe that the formulation for J_c requires $c(x,t)$ to be the velocity associated with directed motion.

Autocorrelation in Movement Direction

The issue of serial autocorrelation in the successive movement directions of individuals has been addressed in a number of earlier papers, including Goldstein (1951), Kareiva and Shigesada (1983), Othmer et al. (1988), Holmes (1993), Turchin (1998), Hillen and Othmer (2000), Hillen and Stevens (2000), Okubo and Grunbaum (2001), Othmer and Hillen (2002). Below we give a brief overview of this issue and refer the reader to the above sources for more details on this subject. Incorporating directional persistence begins with a significant change in the way in which we characterize the underlying movement process. Rather than considering changes in the position and direction of an individual as a function of time, as we did in chapter 2, we switch to a moving frame of reference and consider changes in an individual's speed, direction, and duration of movement. In other words, we switch from viewing movement as a "position jump" process to viewing movement as a "velocity jump" process (Othmer et al. 1988).

Goldstein (1951) analyzed a simple model for a correlated random walk in a single space dimension. Soon thereafter, Patlak (1953) derived the following equation for an individual or particle moving in two space dimensions that exhibits directional persistence in its distribution of movement directions:

$$\frac{\partial u}{\partial t} = \frac{1}{4} \nabla \left[\frac{1 + \psi \left(2 \frac{M_1}{M_2} - 1 \right)}{1 - \psi} \cdot \nabla \left(u \frac{M_2}{\bar{\tau}} \right) \right]$$

$$- \frac{1}{2} \nabla \left[\left(\frac{\mathbf{M}_1}{\tau} + \frac{M_1}{\bar{\tau}} \left([1 + \psi] \mathbf{B} + \frac{\psi}{1 + \psi} \frac{M_1^2}{M_2} \nabla \left[\frac{M_2}{M_1} \right] \right) \right) u \right], \quad \text{(C1)}$$

where the parameter ψ indicates the degree of directional persistence and the parameters M_1, M_2, $\mathbf{M}_1$, $\mathbf{B}$, and $\bar{\tau}$ reflect various properties of the individual's fine-scale movement behavior, specified in terms of a joint distribution of movement speeds and movement durations. Specifically,

$$M_1 = \int \int c\tau f(c, \tau, \mathbf{x}, t)dcd\tau,$$

$$M_2 = \int \int c^2\tau^2 f(c, \tau, \mathbf{x}, t)dcd\tau,$$

$$\tau = \int \int \tau f(c, \tau, \mathbf{x}, t)dcd\tau, \qquad \text{and}$$

$$\mathbf{M}_1 = \left[\int \int \tau f_x(c, \tau, \mathbf{x}, t)dcd\tau, \int \int \tau f_y(c, \tau, \mathbf{x}, t) \right]$$

$$\mathbf{B} = \left[\epsilon_x, \epsilon_y \right], \tag{C2}$$

where $f(c, \tau, \mathbf{x}, t)$ is the joint distribution of the individual's movement speeds c and movement durations τ as a function of its position $\mathbf{x}$ and time t. ϵ_x and ϵ_y are the coefficients of the bias in the x and y directions respectively and f_x and f_y are the marginal distributions of $f(c, \tau, \mathbf{x}, t)$ in the x and y directions respectively. In deriving the above equation, Patlak assumed that the terms governing the bias in movement direction per step length ϵ_x and ϵ_y are small. Formally, the persistence parameter ψ is the mean cosine of the distribution of the individual's turning angles around its current direction of movement. It thus varies between 1 and -1, with values of ψ near 1 indicating strong directional persistence, $\psi = 0$ indicating no directional persistence, and values near $\psi = -1$ indicating a strong propensity to reverse direction at each successive move. Note that equation (C1) is an advection-diffusion equation similar to equation (2.28); however, both the advection and diffusion coefficients are considerably more complex. Setting $\psi = 0$ in equation (C1), we obtain an equation identical to equation (2.28). More recently, Othmer and Hillen (Othmer et al. 1988; Hillen and Othmer 2000; Othmer and Hillen 2002) have performed a more general analysis of velocity jump processes, including an analysis of the conditions necessary to arrive at equation (C1).

APPENDIX D

Estimating the Parameters
of the Localizing Tendency Model

This appendix describes the procedures used in chapter 3 to estimate the parameters of the localizing tendency home range model from high frequency telemetry measurements such as Siniff and Jessan's (1969) red fox dataset shown in figure 3.4. The redistribution kernel underlying the localizing tendency model considered in chapter 3 describes an individual's fine-scale movement behavior (eq. (3.3)) in terms of two component distributions: the individual's distribution of movement directions, given by a von Mises distribution, and the individual's distribution of movement distances $f(\rho)$, given by an exponential distribution.

The von Mises distribution (eq. (3.2)) contains two parameters κ and $\widehat{\phi}$ that respectively determine the magnitude and direction of the bias in movement direction (see figures 3.2 and 3.3). In the case of the localizing tendency model, the direction of the bias is toward the individual's home range center $\vec{x} = [x_H, y_H]$. For the red fox dataset, this was assumed to be the centroid of the relocations, i.e.,

$$x_H = \frac{1}{n_r} \sum_i^{n_r} x_i, \qquad y_H = \frac{1}{n_r} \sum_i^{n_r} y_i, \tag{D1}$$

where x_i and y_i are the x and y coordinates of relocation i and n_r is the number of relocations in the dataset.

κ is calculated from the observed distribution of movement directions ϕ_i relative to the direction of the home range center from the individual's current position $\widehat{\phi}_i$. ϕ_i and $\widehat{\phi}_i$ are given by

$$\phi_i = \tan^{-1} \frac{y_{i+1} - y_i}{x_{i+1} - x_i}$$

$$\widehat{\phi}_i = \tan^{-1} \frac{y_H - y_i}{x_H - x_i}. \tag{D2}$$

From these we can calculate the magnitude of the mean direction vector $\bar{R}$. Defining $\bar{C}$ and $\bar{S}$ as the mean cosine and mean sine of the movement directions relative to the home range center:

$$\bar{C} = \frac{1}{n_r} \sum_{i}^{n_r-1} \cos(\phi_i - \widehat{\phi_i}), \qquad \bar{S} = \frac{1}{n_r} \sum_{i}^{n_r-1} \sin(\phi_i - \widehat{\phi_i}). \qquad \text{(D3)}$$

The magnitude of mean direction vector $\bar{R}$ is given by

$$\bar{R} = \sqrt{\bar{C}^2 + \bar{S}^2}, \qquad \text{(D4)}$$

and κ is related to the mean direction vector by the following equation

$$\bar{R} = \frac{I_1(\kappa)}{I_0(\kappa)}, \qquad \text{(D5)}$$

where I_0 and I_1 are modified Bessel functions. Rather than having to solve equation (D5), it is possible to obtain our estimate for κ, $\widehat{\kappa}$ by using the following approximations:

$$\widehat{\kappa} = 2\bar{R} + \bar{R}^3 + 5\bar{R}^5/6 \qquad\qquad \bar{R} < 0.53$$
$$\widehat{\kappa} = -0.4 + 1.39\bar{R} + 0.43/(1 - \bar{R}) \quad 0.53 \leq \bar{R} < 0.85$$
$$\widehat{\kappa} = 1/(\bar{R}^3 - 4\bar{R}^2 + 3\bar{R}) \qquad\qquad \bar{R} \geq 0.85 \qquad \text{(D6)}$$

(see Fisher 1993).

As shown in figure 3.6, the distribution of distances between successive relocations $(f_\tau(\rho))$ is approximately exponential, i.e.,

$$f_\tau(\rho) \sim \lambda \exp(-\lambda\rho), \qquad \text{(D7)}$$

where $1/\lambda$ is the mean distance between relocations $(\bar{\rho})$. $\bar{\rho}$ was calculated directly from the dataset:

$$\bar{\rho} = \frac{1}{n_r - 1} \sum_{i=1}^{n_r-1} \sqrt{(x_{i+1} - x_i)^2 + (y_{i+1} - y_i)^2}. \qquad \text{(D8)}$$

Movement with Attraction
toward a Den

In this appendix we consider a random walk with attraction toward a den site, and movement described by equation (3.3). We demonstrate how to calculate the advection (eq. (2.14)) and diffusion (eq. (2.15)) terms from the movement kernel $f(r)$ and the turning kernel $K(\phi, \widehat{\phi})$ given in equation (3.3). As an example we consider the case where the turning kernel $K(\phi, \widehat{\phi})$ is described by a von Mises distribution (Batschelet 1981). Without loss of generality, we consider the case where the den site is situated at the origin $(0, 0)$.

Preliminary Analysis

Recalling the definitions of chapter 3, section 3.1, the locations of the individual before and after a jump are given by $\mathbf{x}' = (x', y')$ and $\mathbf{x} = (x, y)$. The distance between the two points is given by $\rho = |\mathbf{x}' - \mathbf{x}|$, the angle of the jump between the two points is given by $\phi = \tan^{-1}(y - y'/x - x')$, and the angle between point $\mathbf{x}$ and the den site, which is assumed to be at the origin, is $\widehat{\phi} = \tan^{-1}(y/x)$. From these definitions we deduce that

$$\cos(\widehat{\phi}) = \frac{x}{|\mathbf{x}|}, \quad \sin(\widehat{\phi}) = \frac{y}{|\mathbf{x}|}, \quad \cos(\phi) = \frac{x - x'}{|\mathbf{x} - \mathbf{x}'|},$$

$$\sin(\phi) = \frac{y - y'}{|\mathbf{x} - \mathbf{x}'|}, \quad \cos(\phi - \widehat{\phi}) = \frac{\mathbf{x}}{|\mathbf{x}|} \cdot \frac{\mathbf{x} - \mathbf{x}'}{|\mathbf{x} - \mathbf{x}'|},$$

$$\sin(\phi - \widehat{\phi}) = \frac{(-y, x)^T}{|\mathbf{x}|} \cdot \frac{\mathbf{x} - \mathbf{x}'}{|\mathbf{x} - \mathbf{x}'|}. \tag{E1}$$

The last two expressions can be derived from the first four, using the double angle formulas for sine and cosine. The formula for $\cos(\phi - \widehat{\phi})$ can be interpreted geometrically as the projection of a unit vector pointing in direction ϕ, namely $(x/|x|)$, onto a unit vector pointing in direction $\phi - \widehat{\phi}$, namely $((\mathbf{x} - \mathbf{x}')/(|\mathbf{x} - \mathbf{x}'|))$. The formula for $\sin(\phi - \widehat{\phi})$ can be interpreted geometrically as the projection of a unit vector pointing in a direction perpendicular to ϕ, namely $((-y, x)^T/|\mathbf{x}|)$, onto a unit vector pointing in direction $\phi - \widehat{\phi}$, namely $((\mathbf{x} - \mathbf{x}')/(|\mathbf{x} - \mathbf{x}'|))$. We use these geometric interpretations when analyzing the advection coefficient below.

We now consider the case where the turning kernel K is an even function of $\theta = \phi - \widehat{\phi}$. We start by evaluating $\mathbf{c}$, d_{xx}, d_{xy}, and d_{yy} from equations (2.14) and (2.15). To start, we observe that the movement rules do not explicitly change with time, so time dependency is dropped in equations (2.14) and (2.15) to yield:

$$\mathbf{c}(\mathbf{x}) = \lim_{\tau \to 0} \frac{1}{\tau} \int_0^{2\pi} \int_0^\infty (\mathbf{x}' - \mathbf{x}) f(\rho) K(\phi - \widehat{\phi}) \, d\rho \, d\phi \qquad \text{(E2)}$$

and

$$d_{xx}(\mathbf{x}) = \lim_{\tau \to 0} \frac{1}{2\tau} \int_0^{2\pi} \int_0^\infty (x' - x)^2 f(\rho) K(\phi - \widehat{\phi}) \, d\rho \, d\phi$$

$$d_{xy}(\mathbf{x}) = d_{xy}(\mathbf{x}) = \lim_{\tau \to 0} \frac{1}{2\tau} \int_0^{2\pi} \int_0^\infty (x' - x)(y' - y) f(\rho) K(\phi - \widehat{\phi}) \, d\rho \, d\phi$$

$$d_{yy}(\mathbf{x}) = \lim_{\tau \to 0} \frac{1}{2\tau} \int_0^{2\pi} \int_0^\infty (y' - y)^2 f(\rho) K(\phi - \widehat{\phi}) \, d\rho \, d\phi. \qquad \text{(E3)}$$

Taking the vector dot product of equation (E2) with the vector pointing toward the den site, $-\mathbf{x}/|\mathbf{x}|$, and employing the trigonometric identities (eq. (E1)) yields

$$-\mathbf{c}(\mathbf{x}) \cdot \frac{\mathbf{x}}{|\mathbf{x}|} = \lim_{\tau \to 0} \frac{1}{\tau} \int_0^\infty \rho f(\rho) \, d\rho \int_0^{2\pi} \cos(\phi - \widehat{\phi}) K(\phi - \widehat{\phi}) \, d\phi. \qquad \text{(E4)}$$

Similarly, multiplying equation (E2) by $(y, -x)^T/|\mathbf{x}|$, a vector perpendicular to $-\mathbf{x}/|\mathbf{x}|$, and employing the trigonometric identities equation (E1) yields

$$\mathbf{c}(\mathbf{x}) \cdot \frac{(y, -x)^T}{|\mathbf{x}|} = \lim_{\tau \to 0} \frac{1}{\tau} \int_0^\infty \rho f(\rho) \, d\rho \int_0^{2\pi} \sin(\phi - \widehat{\phi}) K(\phi - \widehat{\phi}) \, d\phi. \qquad \text{(E5)}$$

Thus our assumption that the turning kernel K is even means that the last integral of equation (E5) equals zero and hence $\mathbf{c}(\mathbf{x}) \cdot (y, -x)^T/|\mathbf{x}| = \mathbf{0}$. No advection in the coordinate perpendicular to the direction of the den site means vector $\mathbf{c}(\mathbf{x})$ points directly toward the den site. The magnitude of the advection term $c = -\mathbf{c}(\mathbf{x}) \cdot \mathbf{x}/|\mathbf{x}|$. The formula for this speed is given by equation (E4).

Using similar methods, the diffusion coefficients (eq. (E3)) can be calculated as

$$d_{xx}(\mathbf{x}) = \lim_{\tau \to 0} \frac{1}{2\tau} \int_0^\infty \rho^2 f(\rho)\, d\rho \int_0^{2\pi} \cos^2(\phi) K(\phi - \widehat{\phi})\, d\phi$$

$$= \lim_{\tau \to 0} \frac{1}{2\tau} \int_0^\infty \rho^2 f(\rho)\, d\rho \left(\frac{1}{2} + \frac{\cos(2\widehat{\phi})}{2} \int_0^{2\pi} \cos(2(\phi - \widehat{\phi})) K(\phi - \widehat{\phi})\, d\phi \right)$$

$$= \lim_{\tau \to 0} \frac{1}{4\tau} \int_0^\infty \rho^2 f(\rho)\, d\rho \left(1 + \frac{x_1^2 - x_2^2}{|\mathbf{x}|} \int_0^{2\pi} \cos(2(\phi - \widehat{\phi})) K(\phi - \widehat{\phi})\, d\phi \right)$$

$$d_{xy}(\mathbf{x}) = d_{yx}(\mathbf{x}) = 0$$

$$d_{yy}(\mathbf{x}) = \lim_{\tau \to 0} \frac{1}{2\tau} \int_0^\infty \rho^2 f(\rho)\, d\rho \int_0^{2\pi} \sin^2(\phi) K(\phi - \widehat{\phi})\, d\phi$$

$$= \lim_{\tau \to 0} \frac{1}{2\tau} \int_0^\infty \rho^2 f(\rho)\, d\rho \left(\frac{1}{2} - \frac{\cos(2\widehat{\phi})}{2} \int_0^{2\pi} \cos(2(\phi - \widehat{\phi})) K(\phi - \widehat{\phi})\, d\phi \right)$$

$$= \lim_{\tau \to 0} \frac{1}{4\tau} \int_0^\infty \rho^2 f(\rho)\, d\rho \left(1 + \frac{x_2^2 - x_1^2}{|\mathbf{x}|} \int_0^{2\pi} \cos(2(\phi - \widehat{\phi})) K(\phi - \widehat{\phi})\, d\phi \right)$$

$$\text{(E6)}$$

Here we have used the fact that K is even, as well as the identity

$$\cos(2\phi) = \cos(2(\phi - \widehat{\phi})) \cos(2\widehat{\phi}) - \sin(2(\phi - \widehat{\phi})) \sin(2\widehat{\phi}),$$

and standard trigonometric identities.

We now discuss properties of the von Mises distribution prior to calculating the advection and diffusion coefficients from the underlying stochastic movement process.

Von Mises Distribution

The following description closely follows Batschelet (1981). The von Mises distribution is a widely used circular distribution that plays a similar role to the normal distribution in linear statistics It is also sometimes referred to as the circular normal distribution (Gumbel 1954).[1]

It is a unimodal distribution with probability density function

$$K(\phi, \widehat{\phi}) = \frac{1}{2\pi I_0(\kappa)} \exp\left[\kappa \cos(\phi - \widehat{\phi}) \right], \tag{E7}$$

[1] Not to be confused with the *wrapped* normal distribution (Batschelet 1981).

with parameter κ ($\kappa \geq 0$), and $I_0(\kappa)$ as a modified Bessel function of the first kind and of zeroth order. The angle $\widehat{\phi}$ is the mode of the distribution and, since the distribution is symmetric about the mode, it is also the mean direction. The parameter κ is the concentration parameter that governs the degree of non-uniformity in the distribution of direction (see figure 3.2). When $\kappa = 0$, the distribution becomes a circular uniform distribution. The expected value of ϕ is $\widehat{\phi}$ and the expected value of $\cos(n(\phi - \widehat{\phi}))$ is $I_n(\kappa)/I_0(\kappa)$. When $n = 1$, the center of mass is given by the mean vector, which points in the direction $\widehat{\phi}$, and whose magnitude m is a monotonically increasing function of the concentration parameter κ:

$$m(\kappa) = I_1(\kappa)/I_0(\kappa). \tag{E8}$$

From von Mises to Advection-Diffusion

We now evaluate equation (E4) and eq. (E6) with the von Mises distribution equation (E7) used for K. Because the advection-diffusion limit requires the infinitesimal first and second moments of the redistribution kernel k, we assume that both the distribution of distances f and the distribution of turning angles K depend implicitly upon the length of the time step τ. We define

$$m_f^1(\tau) = \int_0^\infty \rho f(\rho) \, d\rho \tag{E9}$$

and

$$m_f^2(\tau) = \int_0^\infty \rho^2 f(\rho) \, d\rho \tag{E10}$$

and the parameter $\kappa(\tau)$ in the von Mises distribution to also depend on τ so that each of these quantities approaches zero as τ approaches zero. We use the series expansions for the modified Bessel functions

$$I_0(\kappa) = 1 + \frac{1}{4}\kappa^2 + \text{h.o.t.}, \tag{E11}$$

$$I_1(\kappa) = \frac{1}{2}\kappa + \text{h.o.t.}, \tag{E12}$$

$$I_2(\kappa) = \frac{1}{8}\kappa^2 + \text{h.o.t.}, \tag{E13}$$

and retain terms to leading order to evaluate equation (E4) as

$$c = \lim_{\tau \to 0} \frac{m_f^1(\tau)\kappa(\tau)}{2\tau}, \tag{E14}$$

and equation (E6) as

$$d_{xx} = d_{yy} = d = \lim_{\tau \to 0} \frac{m_f^2(\tau)}{4\tau}. \tag{E15}$$

Here we have assumed that the mean displacement distance $m_f^1(\tau)$, the mean-squared displacement distance $m_f^2(\tau)$, and the concentration parameter $\kappa(\tau)$ all approach zero as τ approaches zero and that, in doing this, they scale so that equation (E14) and equation (E15) remain well defined. This is the typical assumption that is made when taking the diffusion limit of a random walk to yield a advection-diffusion equation (Okubo 1980).

Model Fitting

In several chapters in this book, we used the method of maximum likelihood to fit mechanistic home range models to the relocation datasets. Here we provide some further details on this method. The likelihood (L) of a particular hypothesis (H) given a dataset (R) is proportional to the probability of obtaining the data given the hypothesis, i.e.,

$$L(H|R) = kP(R|H), \tag{F1}$$

where k is an arbitrary constant. In the context of fitting mechanistic home range models to empirical home range data, R is a dataset for spatial locations of the relocations (i.e., $R = [x_j, y_j]$ $j = 1 \ldots n_r$, where n_r is the number of relocations), and the hypothesis, H, is the mechanistic home range model with a specific set of parameter values.

The probability of observing an individual at a particular location given an underlying model of movement is proportional to the expected space use predicted by the home range model at that location $u(x, y)$. Thus, provided the relocations are sufficiently spaced that they can be considered spatially independent (the meaning of independence in this context is discussed in more detail in chapter 10), the likelihood of obtaining the observed set of relocations R is

$$L = k \prod_{j=1}^{n_r} u(x_j, y_j), \tag{F2}$$

The maximum likelihood estimate for a given movement model is the set of model parameters that give the maximum possible value for L. Since the logarithm is an increasing function, in practice we maximize the log-likelihood:

$$l(\theta) = k + \sum_{j=1}^{n_r} u(x_j, y_j), \tag{F3}$$

where $l = \ln(L)$ and θ indicates the set of model parameters whose values are to be maximized. k is now simply an additive constant whose value does not change with the value of the parameters we are trying to estimate and thus can be dropped from the above equation.

In chapter 3, where we fit the Holgate-Okubo mechanistic home range model to the relocations of the Hopsage pack at Hanford, $u(x_j, y_j)$ is the space use at position $(x_j \ y_j)$ given by the solution of equation (3.5), and $\theta = \beta$, the model's parameter whose value is to be maximized. The location of the home center $\vec{x}$ is assumed to be the centroid of the relocations. We determine the value of β that maximizes the value of equation (F3). In this case, where we have an analytic expression for the steady-state pattern of space use (eq. 3.14), the maximization can be done analytically, by differentiating the log-likelihood function (eq. F3) with respect to β, and solving for the critical point $dL/d\beta = 0$. This yields the following equation for the maximum likelihood estimate of β:

$$\widehat{\beta}^2 \sum_{j=1}^{n_r} \frac{\exp(-\widehat{\beta}r_j)}{u(r_j)} = 2\pi n_r \tag{F4}$$

where $u(r_j)$ is given by equation (3.14) and r_j is the radial distance of point (x_j, y_j) from the home range center $\vec{x}$.

The other mechanistic home range models developed in this book have steady-state space use equations that cannot be solved analytically and contain multiple parameters whose values need to be simultaneously maximized. For example, the fitting of the conspecific avoidance home range model to the relocation data for the six packs at Hanford in chapter 4 requires that we maximize the following equation:

$$l(\beta, m) = \sum_{i=1}^{n} \sum_{j=1}^{n_{r_i}} \ln u^{(i)}(x_{ij}, y_{ij}), \tag{F5}$$

with respect to the two parameters of this model, β and m, where $u^{(i)}(x_{ij}, y_{ij})$ is the height of the probability density function (pdf) for expected space use by pack i at point (x_{ij}, y_{ij}) given by the steady-state solution of equations (4.20)–(4.21), and (x_{ij}, y_{ij}) are the spatial coordinates of relocations for individuals belonging to pack i ($i = 1 \ldots n$, where n is the number of packs and $j = 1 \ldots n_{r_i}$, where n_{r_i} is the total number of relocations for pack i).

We maximize value of the log-likelihood equation (eq. (F5)) with respect to β and m using a numerical maximization algorithm, solving equations (4.20)–(4.21) numerically for each set of parameter values. For further information on numerical maximization methods, see Press et al. (1992a) or Press et al. (1992b). More details on numerical methods for solving the partial differential equations can be found in appendix G. For further details on the method of maximum likelihood, see Edwards (1992).

Numerical Methods for Solving
Space Use Equations

For the biologically realistic case of home ranges in two-dimensional space, the conspecific avoidance model equations (eqs. (4.20) and (4.21)) could not be solved analytically and so were solved by numerical simulation. We used the "method of lines" (Schiesser 1991) to approximate the PDEs (eq. (4.20)) by discretizing the spatial derivative terms to second order, yielding a large set of spatially coupled ODEs. The scent-marking ODEs (eq. (4.21)) were discretized onto the same spatial grid used to discretize the associated PDEs. Using this method, all the models could be represented by large systems of coupled ODEs.

The number of ODEs used to represent the equations is governed by the size of the simulation region and the spatial scale used for discretization. We simulated the model equations in a 12.5 km × 11.0 km domain encompassing the relocation data, at a spatial resolution of 100 m × 100 m, since fine-scale approximations of the spatial derivatives were necessary for accurate simulation of the PDEs. Due to the flux-conserving nature of equation (4.20) with zero-flux boundary conditions (eq. (4.22); see also Press et al. 1992), we solved for the steady-state of PDEs (eq. (4.20)) and ODEs (eq. (4.21)) by solving the corresponding time-dependent equations. For example, with the two-pack case, the time-dependent equations corresponding to equations (4.15)–(4.18) are given in equations (4.11)–(4.14).

From a prescribed initial condition of uniform space use across the study area by all packs, we iterated the equations to convergence, a procedure known as the "method of false transients" (Ames 1992). The time integration was done using a fully implicit method to solve the system of coupled equations at each time step (Saad and Schultz 1986; Brown et al. 1989; Byrne 1992). Further details of the simulation and fitting methods can be found in Moorcroft (1997).

Displacement Distances

The first part of this appendix gives details of the calculations from chapter 10. The second part calculates the furthest distance an individual has moved from the den site as a function of time.

Furthest Distance Sampled

Given a probability density function $p(s)$ $(0 \leq s < \infty)$, the expected value of the random variable s is

$$\bar{s} = \int_0^\infty sp(s)\,ds. \tag{H1}$$

Defining $P(s)$ as $\int_0^s p(y)dy$, the cumulative density function of $p(s)$, integration by parts yields

$$\bar{s} = \int_0^\infty sP'(s)\,ds$$

$$= [s(P(s) - 1)]_0^\infty + \int_0^\infty (1 - P(s))\,ds$$

$$= \int_0^\infty (1 - P(s))\,ds. \tag{H2}$$

Substituting equation (10.3) into equation (H2), then using the binomial expansion theorem to expand the terms, yields

$$\bar{s} = \int_0^\infty \left(1 - \sum_{k=0}^{n_r} \binom{n_r}{k}(-1)^k \exp(-\beta sk)\right) ds. \tag{H3}$$

Integrating the sum term-wise yields equation (10.5).

A similar method is used to calculate the expected maximum displacement distance for the truncated dataset. Equations (10.7) and (H2) yield

$$\bar{s}_{tr} = \bar{s} - \sum_{j=1}^{m} \binom{n_r}{j} \int_{0}^{\infty} (1 - \exp(-\beta s))^{n_r - j} \exp(-\beta s j) \, ds. \qquad (H4)$$

Application of the binomial expansion theorem and integration term-wise yields equation (10.8).

Furthest Distance Moved

If individuals undergo a random walk with simple bias toward the den site, then they may stray a long distance away from the den site, due to a series of random steps, before eventually returning. We can be sure to capture any major excursion from the den site by ensuring movements are sampled over shorter and shorter time intervals. This kind of short time-interval sampling is now possible using new radio-tracking methods that can record locations every few minutes. In this situation, the analysis of the furthest sampled distance from the den site in chapter 10 (see section 10.1) is not applicable because the observations are not spatially independent. Specifically, the location of the individual over short time intervals depends on its current location. A new approach is needed. Here it is most useful to return to the original partial differential equation (eq. (10.1)).

If the individual undergoes a random walk with simple bias toward the den site, as described by equation (10.1), then what is the probability density function for the furthest distance from the den site achieved in a fixed time interval, and how does this distance change with the length of the time interval? We begin by considering a related problem. Suppose an individual moves according to equation (10.1) on a domain $-y < x < y$ with absorbing boundary conditions at $\pm y$. Here the absorbing boundary conditions $u(-y, t) = u(y, t) = 0$ describe removal of the individual as soon as it touches points $x = \pm y$. Given an initial probability density function for the individual $u_0(x)$ defined upon $-y < x < y$, the probability that the individual has not yet passed beyond y at time t is given by the area remaining under the solution curve $\int_{-y}^{y} u(x, t) \, dx$. Initially this area is equal to 1, and it decreases with time due to the increasing chance that the individual has moved beyond $\pm y$.

We can repeat this procedure for all possible distances y, and define the solution u, parameterized by maximum possible distance traveled before removal y, as $u(x, t; y)$. Then, given the initial probability density function for the individual $u_0(x)$, the probability that the maximum distance attained by the individual by time t is less than y is

$$P(y,t) = \int_{-y}^{y} u(x,t;y)\, dx, \tag{H5}$$

and the related probability density function for the furthest displacement y up to time t is $p(y,t) = \partial P/\partial y$. This approach can be generalized to the radially symmetric problem (eq. (3.12)).

While the numerical analysis of the above problem involves solving a partial differential equation with new boundary conditions for each possible y, the mathematical analysis using separation of variables allows for an explicitly calculated solution which is valid for each y, c, and d. By way of example, we consider the case where the individual is released at location $x = 0$ at time $t = 0$ so that $u_0(x) = \delta(x)$. Spatial symmetry in the initial distribution and in equation (10.1) allow us to rewrite the model in terms of distance to the den site $\xi = |x|$. The model is

$$\frac{\partial u}{\partial t} = c\frac{\partial u}{\partial \xi} + d\frac{\partial^2 u}{\partial \xi^2} \tag{H6}$$

with zero-flux boundary conditions $d\partial u/\partial \xi + cu = 0$ at $\xi = 0$ arising from the solution symmetry, zero boundary conditions $u = 0$ at $\xi = y$ arising from the absorbing condition, and the initial condition now given as $u_0(\xi) = \delta(\xi)$.

The method of separation of variables can be used to solve this equation to give

$$u(\xi,t;y) = 2\exp(-c\xi/(2d) - c^2t/(4d))$$
$$\sum_{i=1}^{\infty} \exp(-d\lambda_n^2 t)\frac{\sin(\lambda_n(y-\xi))\sin(\lambda_n y)}{y - \sin(\lambda_n y)\cos(\lambda_n y)/\lambda_n}, \tag{H7}$$

where the eigenvalue λ_n is the nth positive solution satisfying

$$\lambda_n = c\tan(\lambda_n y)/(2d). \tag{H8}$$

Figure H.1 shows a sample solution $u(\xi,t;y)$ as a function of distance (ξ) for a fixed value of y ($y = 1$) and varying t. The absorbing boundary condition at $x = y = 1$ causes the area under the curve $u(\xi,t;y)$ to decrease with time. The cumulative density function, describing the probability that the maximum distance attained by an individual by time t is less than y, is given by

$$P(y,t) = \int_{0}^{y} u(\xi,t;y)\, dx = 2\exp(-c^2t/4d) \tag{H9}$$
$$\sum_{i=1}^{\infty} \frac{\lambda_n \exp(-d\lambda_n^2 t)\sin(\lambda_n y)\exp(-cy/(2d))}{(y - \sin(\lambda_n y)\cos(\lambda_n y)/\lambda_n)(c^2/4d^2 + \lambda_n^2)}.$$

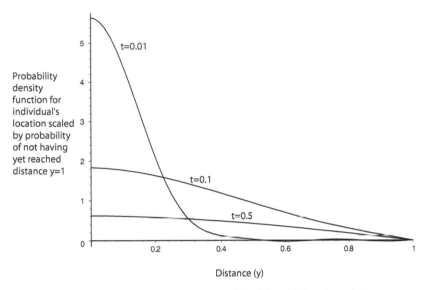

FIGURE H.1. Solution to equation (H7) for $t = 0.01, 0.1,$ and $0.5,$ and $y = 1.$ Parameter values are $c = 0.1$ and $d = 1.$

Figure H.2 shows the cumulative density function $P(y, t)$ for a variety of fixed time values t and varying distances y. The decrease in area under the curves in figure H.1 with increasing time are reflected in the curves shifting to the right with increasing time in figure H.2.

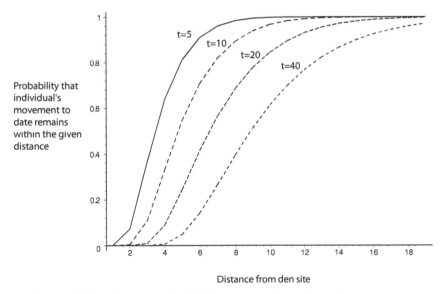

FIGURE H.2. Solution to equation (H9) for various values of time t. Parameter values are $c = 0.1$ and $d = 1.$

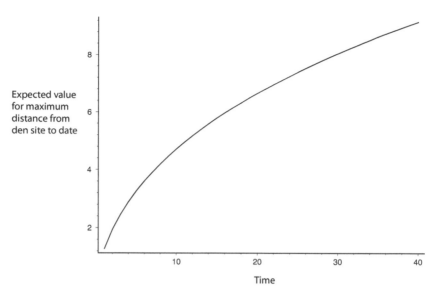

FIGURE H.3. The expected value for the furthest distance traveled to date (calculated with eqs. (H9) and (H2)). Parameter values are $c = 0.1$ and $d = 1$.

Finally, the expected value of the maximum distance attained by an individual to date can be calculated from the cumulative density function (eq. (H9)) for different time values. Figure H.3 shows the expected value of the maximum distance attained by an individual as a function of time t. The expected value of the maximum distance attained by an individual to date increases monotonically with time, even though there is a term biasing movement back toward the den site. Here, the longer the measurement time interval, the more opportunity the individual has to roam widely. Unlike the furthest distance *measured* over repeated independent samples, which depends only upon the bias $\beta = c/d$, the furthest distance *traveled* over a time interval depends independently upon each of directed motion term (c) and the random motion term (d).

ESS Analysis Model Parameters

The growth ratio λ in equations (9.1)–(9.5) was chosen as $\lambda = 2$, giving recruitment rates consistent with empirical estimates of 30% per annum (Nelson and Mech, 1981) at approximately 60% of carrying capacity (the typical equilibrium population level; for example, see figure 11.3). The predation rate was estimated assuming that deer have a mean lifespan of approximately seven years, consistent with expirial estimates (Nelson and Mech, 1981), and giving individual deer a 10–15% chance of being killed each year. The precise probability upon location of the deer in relation to the wolf home ranges. For the case where the interactions and "Pure Territorial" and symmetric (eq. (11.16)), the predation pressure is uniform across the region $(u(x) + v(x) = 2)$. Here, a choice of $\psi = 0.15$ yields the probability of surviving a year as

$$\frac{1 - exp(-\psi(u(x) + v(x)))}{(\psi(u(x) + v(x)))} \approx 0.86 \tag{I1}$$

and thus a 14% chance of being killed each year.

References

Adams, E. (2001). Approaches to the study of territory size and shape. *Annual Review of Ecology and Systematics 32*, 277–303.

Adler, F. R., and D. M. Gordon (2003). Optimization, conflict, and nonoverlapping foraging ranges in ants. *American Naturalist 162*, 529–543.

Aebischer, N. J., Robertson, P. A., and Kenward, R. E. 1993. Compositional analysis of habitat use from animal radio-tracking data. *Ecology 74*, 1313–1325.

Ahearn, S., J. Smith, and A. Joshi (2001). TIGMOD: An individual-based spatially explicit model for simulating tiger/human interaction in multiple use forest. *Ecological Modelling 140*(1–2), 81–97.

Alldredge, J. R., and Ratti, J. T. (1986). Comparison of some statistical techniques for analysis of resource selection. *J. Wildl. Manage. 50*, 157–165.

Alt, W. (1980). Biased random walk models for chemotaxis and related diffusion approximations. *Journal of Mathematical Biology 9*, 147–177.

Ames, W. (1992). *Numerical methods for partial differential equations*. Boston: Academic Press.

Anderson, A. R. A., and M. A. J. Chaplain (1998). Continuous and discrete mathematical models of tumor-induced angiogenesis. *Bulletin of Mathematical Biology 60*, 857–900.

Arthur, S. M., Manly, B. F. J., McDonald, L. L., and Garner, G. W. (1996). Assessing habitat selection when availability changes. *Ecology 77*, 215–227.

Ballard, W. B., L. Ayres, P. Kruasman, D. Reed, and S. Fancy (1998). Ecology of wolves in relation to a migratory caribou herd in northwest Alaska. *Wildlife Monographs 135*, 1–47.

Ballard, W. B., L. N. Carbyn, and D. W. Smith (2003). Wolf interactions with non-prey. In L. D. Mech and L. Boitani (Eds.), *Wolves: Behavior, ecology and conservation*, pp. 259–271. Chicago: University of Chicago Press.

Ballard, W. B., J. Whitman, and C. Gardner (1987). Ecology of an exploited wolf population in south-central Alaska. *Wildlife Monographs 98*, 1–54.

Barrett, L., and C. Lowen (1988). Random walks and the gas model: Spacing behaviour of grey-cheeked mangabey. *Functional Ecology 12*, 857–865.

Barrette, C., and F. Messier (1980). Scent marking in free ranging coyotes. *Animal Behavior 28*, 814–819.

Batschelet, E. (1981). *Circular statistics in biology*. Toronto: Academic Press.

Bekoff, M., and T. Daniels (1984). Life history patterns and comparative social ecology of carnivores. *Annu.Rev.Ecol.Syst. 15*, 191–232.

Bekoff, M., and M. Wells (1980). The social ecology of coyotes. *Scientific American 242*, 130–151.

Bekoff, M., and M. Wells (1981). Behavioral budgeting by wild coyotes: The influence of food resources and social organization. *Animal Behavior 29*, 794–801.

Bekoff, M., and M. Wells (1982). Behavioral ecology of coyotes. *Zeitschrift für Tierpsychologie 60*, 281–305.

Bekoff, M., and M. Wells (1986). The social ecology of coyotes. In J. Rosenblatt (Ed.), *Advances in the study of behavior*, pp. 251–338. New York: Academic Press.

Benhamou, S. (1988). Analysing spatial relationships from trapping data. *Journal of Mammalogy 69*, 828–831.

Benhamou, S. (1989). An olfactory orientation model for mammals' movements in their home ranges. *Journal Theoretical Biology 139*, 379–388.

Berg, H. C. (1993). *Random walks in biology*. Princeton NJ: Princeton University Press.

Berg, H. C., and D. A. Brown (1974). Chemotaxis in *e. coli* analyzed by three dimensional tracking. *Nature 239*, 500–504.

Bharucha-Reid, A. (1960). *Elements of the theory of Markov processes and their applications*. New York: McGraw-Hill.

Bowen, W. (1981). Coyote social organization and prey size. *Canadian Journal of Zoology 59*, 639–652.

Bowen, W. (1982). Home range and spatial organization of coyotes in Jasper National Park, Alberta. *Journal of Wildlife Management 46(1)*, 201–216.

Bowen, W., and I. McTaggart-Cowan (1980). Scent marking in coyotes. *Canadian Journal of Zoology 58*, 639–652.

Boyce, M. S., J. Mao, E. Merril, D. Fortin, M. Turner, J. Fryxell, and P. Turchin (2004). Scale and heterogeneity in habitat selection by elk in Yellowstone National Park. *Ecoscience 10*, 234–245.

Briscoe, B., M. A. Lewis, and S. Parrish (2002). Home range formation in wolves. *Bulletin of Mathematical Biology 64*, 261–284.

Brown, P., G. Byrne, and A. Hindmarsh (1989). VODE: A variable-coefficient ODE solver. *SIAM J. Sci. Stat. Comput. 10*, 1038–1051.

Burnham, K. P., and D. R. Anderson (2002). *Model selection and multimodal inference: An information-theoretic approach*. New York: Springer Verlag.

Byrne, G. (1992). Pragmatic experiments with Krylov methods in the stiff ODE setting. In J. Cash and I. Gladwell (Eds.), *Computational ordinary differential equations*, pp. 323–356. Oxford: Oxford University Press.

Calhoun, J., and J. Casby (1958). *Calculation of home range and density of small mammals*, volume 55. Washington, DC: U.S. Government Printing Office.

Camenzind, F. (1978). Behavioral ecology of coyotes on the national elk refuge. In M. Bekoff (Ed.), *Coyotes: Biology, behavior and management*. New York: Academic Press.

Carbyn, L. (1980). Ecology and management of wolves in Riding Mountain National Park, Manitoba. Canadian Wildlife Service. Technical report 10, Edmonton, Alberta.

Clutton-Brock, T. (1989). Mammalian mating systems. *Proceedings of the Royal Society 236*, 339–372.

Clutton-Brock, T., F. Guinness, and S. Albon (1982). *Red Deer. Behavior and Ecology of Two Sexes*. Chicago: Chicago University Press.

Cochran, W., and R. Lord (1963). A radio-tracking system for wild animals. *Journal of Wildlife Management 27*, 9–24.

Comiskey, E., L. J. Gross, D. Fleming, M. Huston, O. Bass, H.-K. Luh, and Y. Wu (1997). A spatially-explicit individual-based simulation model for Florida panther and white-tailed deer in the Everglades and Big Cypress landscapes.

In D. Jordan (Ed.), *Proceedings of the Florida Panther Conference, Ft. Myers Fla., Nov. 1994*, pp. 494–503. U.S. Fish and Wildlife Service.

Conroy, M., Y. Cohen, F. James, Y. Matsinos, and B. A. Maurer (1995). Parameter estimation, reliability, and model improvement for spatially explicit models of animal populations. *Ecological Applications 5*, 17–19.

Cooper, A. B., and J. J. Millspaugh (2001). Accounting for variation in resource availability and animal behavior in resource selection studies. In J. Millspaugh and J. M. Marzluff (Eds.), *Radio tracking and animal populations*, pp. 246–273. San Diego: Academic Press.

Couzin, I. D., and J. Krause (2003). Self-organization and collective behavior in vertebrates. *Advances in the Study of Behavior 32*, 1–75.

Crabtree, R. (1989). *Sociodemography of an unexploited coyote population*. Ph.D. thesis, University of Idaho, Moscow, ID.

Crabtree, R., and J. Sheldon (1999). Coyotes and canid coexistence in Yellowstone. In T. Clark, A. Curlee, S. Minta, and P. Karieva (Eds.), *Carnivores in ecosystems: The Yellowstone experience*, pp. 127–163. New Haven: Yale University Press.

Creel, S., and N. Creel (1996). Limitation of african wild dogs by competition with larger carnivores. *Conservation Biology 10*, 526–538.

Creel, S., G. Spong, and N. Creel (2001). Interspecific competition and the population biology of extinction-prone carnivores. In J. L. Gittleman, S. M. Funk, D. MacDonald, and R. K. Wayne (Eds.), *Conservation biology 5: Carnivore conservation*, pp. 35–60. Cambridge: Cambridge University Press.

DeAngelis, D. L., and L. J. Gross (1992). *Individual-based models and approaches in ecology: populations, communities, and ecosystems*. New York: Chapman and Hall.

Dixon, K., and J. Chapman (1980). Harmonic mean measure of animal activity areas. *Ecology 61*, 1040–1044.

Dunning, J., B. Danielson, B. Noon, T. Root, R. L. Lamberson, and E. Stevens (1995). Spatially explicit population models: Current forms and future uses. *Ecological Applications 5*, 3–11.

Edelstein-Keshet, L. (1988). *Mathematical models in biology*. New York: Random House.

Edwards, A. (1992). *Likelihood*. Baltimore: Johns Hopkins University Press.

Eisenberg, J., and D. Kleiman (1972). Olfactory communication in mammals. *Annual Review of Ecology and Systematics 3*, 1–32.

Erickson, W. P., T. L. McDonald, K. G. Gerow, S. Howlin, and J. W. Kern (2001). Statistical issues in resource selection studies with radio marked animals. In J. Millspaugh and J. M. Marzluff (Eds.), *Radio tracking and animal populations*, pp. 211–242. San Diego: Academic Press.

Fisher, N. I. (1993). *Statistical analysis of circular data*. Cambridge: Cambridge University Press.

French, J., and J. Cleveland (1984). Scent marking in the tamarin (*saguinus oedipus*): Sex differences and ontology. *Animal Behavior 32*, 615–623.

Fritts, S. H., and L. D. Mech (1981). Dynamics, movements, and feeding ecology of a newly protected wolf population in northwestern Minnesota. *Wildlife Monographs 80*, 1–79.

Fuller, T. (1989). Population dynamics of wolves in north-central Minnesota. *Wildlife Monographs 105*, 1–41.

Gese, E., and S. Grothe (1995). Analysis of coyote predation on deer and elk during winter in Yellowstone National Park. *American Midland Naturalist 133*, 36–43.

Gese, E., and R. Ruff (1996). Scent-marking by coyotes (*Canis latrans*): The influence of social and ecological factors. *Animal Behavior 54*, 1155–1166.

Gese, E., R. Ruff, and R. Crabtree (1996a). Foraging ecology of coyotes (*Canis latrans*): The influence of extrinsic factors and a dominance hierarchy. *Canadian Journal of Zoology 74*, 769–783.

Gese, E., R. Ruff, and R. Crabtree (1996b). Intrinsic and extrinsic factors influencing coyote predation of small mammals in Yellowstone National Park. *Canadian Journal of Zoology 74*, 784–797.

Ginsberg, J. (2001). Setting priorities for carnivore conservation: What makes carnivores different? In J. L. Gittleman, S. M. Funk, D. Macdonald, and R. K. Wayne (Eds.), *Carnivore conservation*, pp. 498–523. Cambridge: Cambridge University Press.

Ginsberg, J., and D. Macdonald (1990). *Foxes, wolves, jackals and dogs: an action plan for the conservation of canids.* Gland, Switzerland: IUCN.

Girard, I., J. Ouellet, R. Courtois, C. Dussault, and L. Breton (2002). Effects of sampling effort based on GPS telemetry on home-range size estimations. *Journal of Wildlife Management 66*, 1290–1300.

Gittleman, J. L. (1985). Carnivore body size: Ecological and taxonomic correlates. *Oecologia 67*, 540–554.

Gittleman, J. L., S. M. Funk, D. Macdonald, and R. K. Wayne (2001). *Carnivore conservation*. Cambridge: Cambridge University Press.

Gittleman, J. L., and P. H. Harvey (1982). Carnivore home range size, metabolic needs and ecology. *Behavioral Ecology and Sociobiology 10*, 57–63.

Goldstein, S. (1951). On diffusion from discontinuous movements and on the telegraph equation. *Quarterly Journal of Mechanics and Applied Mathematics 4*, 129–156.

Gorman, M., and M. Mills (1984). Scent marking strategies in hyaenas. *Journal of Zoology 202*, 535–547.

Grunbaum, D. (1998). Using spatially explicit models to characterize foraging performance in heterogeneous landscapes. *American Naturalist 151*(2), 97–115.

Grunbaum, D. (2000). Advection-diffusion equations for internal state-mediated random walks. *SIAM Journal on Applied Mathematics 61*, 43–73.

Gueron, S., and S. A. Levin (1993). Self-organization of front patterns in large wildebeest herds. *J. Theor. Biol. 165*, 541–552.

Gumbel, E. (1954). Applications of the circular normal distribution. *Journal of the American Statistical Association 49*, 267–297.

Haberman, R. (1987). *Elementary applied partial differential equations* (2d ed.). Englewood Cliffs, NJ: Prentice-Hall.

Harrison, D., and J. Gilbert (1985). Denning ecology and movements of coyotes in Maine during pup rearing. *Journal of Mammalogy 66*(4), 712–719.

Hilborn, R., and M. Mangel (1997). *The ecological detective*. Princeton, NJ: Princeton University Press.

Hill, N., and D. Hader (1997). A biased random walk model for the trajectories and swimming micro-organisms. *Journal of Theoretical Biology 186*, 503–526.

Hillen, T., and H. Othmer (2000). The diffusion limit of transport equations derived from velocity jump processes. *SIAM Journal of Applied Mathematics 61*, 751–775.

Hillen, T., and A. Stevens (2000). Hyperbolic models for chemotaxis in 1-D. *Nonlinear Analysis: Real World Applications 1*, 409–433.

Hixon, M. (1980). Food production and competitor density as the determinants of feeding territory size. *American Naturalist 115*, 510–530.

Holgate, P. (1971). Random walk models for animal behavior. In G. Patil, E. Pielou, and W. Walters (Eds.), *Statistical ecology: Sampling and modeling biological populations and population dynamics*, vol. 2 of *Penn State statistics*, pp. 1–12. University Park, PA: Penn State University Press.

Holling, C. (1966). The strategy of building models in complex ecological systems. In K. Watt (Ed.), *Systems analysis in ecology*. New York: Academic Press.

Holmes, E. (1993). Are diffusion models too simple? A comparison with telegraph models of invasion. *American Naturalist 142*, 403–419.

Holzman, S., and M. Conroy (1992). Home range movements and habitat use of coyotes in south-central Georgia. *Journal of Wildlife Management 56*, 139–141.

Hoskinson, R., and L. Mech (1976). White-tailed deer migration and its role in wolf predation. *Journal of Wildlife Management 40*, 429–441.

Jarman, P. (1974). The social organization of antelope in relation to their ecology. *Behaviour 48*, 215–267.

Jennrich, R., and F. Turner (1969). Measurement of non-circular home range. *Journal of Theoretical Biology 22*, 227–237.

Johnson, D. (1980). The comparison of usage and availability measurements for evaluating resource preference. *Ecology 61*, 65–71.

Johnson, R. (1973). Scent marking in mammals. *Animal Behavior 21*, 521–525.

Johnson, W., T. Fuller, and W. Franklin (1996). Sympatry in canids: A review and assessment. In J. Gittleman (Ed.), *Carnivore behavior, ecology and evolution*, vol. 2. Ithaca, NY: Cornell University Press.

Johnstone, R. (1997). The evolution of animal signals. In J. Krebs and N. Davies (Eds.), *Behavioural ecology: An evolutionary approach (4th ed.)*, pp. 155–178. Oxford: Blackwell Scientific Publications.

Jolly, A. (1966). *Lemur behavior: A Madagascar field study*. Chicago: University of Chicago Press.

Kac, M. (1947). Random walk and the theory of Brownian motion. *American Mathematical Monthly 54*, 369–391.

Kareiva, P., and G. Odell (1987). Swarms of predators exhibit preytaxis, if individual predators use over–restricted. *American Naturalist 130*, 233–270.

Kareiva, P. M., and N. Shigesada (1983). Analyzing insect movement as a correlated random walk. *Oecologia 56*, 234–238.

Kelt, D. A., and D. V. Vuren (2001). The ecology and macroecology of mammalian home-range area. *American Naturalist 157*, 637–645.

Kernohan, B. J., R. A. Gitzen, and J. J. Millspaugh (2001). Analysis of animal space-use and movements. In J. Millspaugh and J. M. Marzluff (Eds.), *Radio tracking and animal populations*, pp. 125–166. San Diego: Academic Press.

Kodric-Brown, A., and J. Brown (1978). Influence of economics, inter-specific competition and sexual dimorphism on territoriality of migrant rufous hummingbirds. *Ecology 59*, 285–296.

Krebs, J., and R. Dawkins (1984). Animal signals: Mind reading and manipulation. In J. Krebs and N. Davies (Eds.), *Behavioral ecology: An evolutionary approach* (2d ed.). Oxford: Blackwell Scientific Publications.

Kruuk, H. (1972). *The spotted hyena: A study of predation and social behavior*. Chicago: University of Chicago Press.

Kruuk, H. (1976). Feeding and social behavior of the striped hyaena. *East African Wildlife Journal 14*, 91–111.

Kruuk, H. (1978). Spatial organization and territorial behavior of the European badger *Meles meles*. *Journal of the Zoological Society of London 184*, 1–19.

Laundre, J., and B. Keller (1981). Home range use by coyotes in Idaho. *Animal Behavior 29*, 449–461.

Levin, S., and S. Pacala (1996). Theories of simplification and scaling in ecological systems. In D. Tilman and P. Kareiva (Eds.), *Spatial ecology: The role of space in population dynamics and interspecific interactions*. Princeton, NJ: Princeton University Press.

Levins, R. (1966). The strategy of model building in population biology. *American Scientist 54*, 421–431.

Lewis, M. A., and P. R. Moorcroft (2001). ESS analysis of mechanistic home range models: The value of signals in spatial resource partitioning. *J. Theor. Biol. 210*, 449–461.

Lewis, M. A., and J. D. Murray (1993). Modelling territoriality and wolf-deer interactions. *Nature 366*, 738–740.

Lewis, M. A., K. White, and J. Murray (1997). Analysis of a model for wolf territories. *Journal of Mathematical Biology 35*, 749–774.

Lima, S. L., and P. A. Zollner (1996). Towards a behavioral ecology of ecological landscapes. *Trends in Ecology and Evolution 11*, 131–135.

Macdonald, D. W. (1979). Some observations and field experiments on the urine marking behavior of the red fox *Vulpes vulpes*. *Zeitschrift für Tierpsychologie 51*, 1–22.

Macdonald, D. W. (1980a). Patterns of scent marking with urine and faeces amongst carnivore populations. *Symposium of the Zoological Society of London 45*, 107–139.

Macdonald, D. W. (1980b). The red fox *Vulpes vulpes*, as a predator upon earthworms, *Lumbricus terrestris*. *Zeitschrift für Tierpsychologie 52*, 171–200.

Macdonald, D. W. (1983). The ecology of carnivore social behavior. *Nature 301*, 379–383.

Macdonald, D. W., F. Ball, and N. Hough (1980). The evaluation of home range size and configuration using radio tracking data. In C. Amlaner and D. W. Macdonald (Eds.), *A handbook on biotelemetry and radio tracking*. Oxford: Pergamon Press.

Macdonald, D. W., and P. Moehlman (1973). Cooperation, altruism and restraint in the reproduction of carnivores. In P. Bateson and P. Klopfer (Eds.), *Perspectives in ethology* Vol. 1, pp. 443–467. New York: Plenum Press.

MacDonald, D. W., and S. Rushton (2003). Modelling space use and dispersal of mammals in real landscapes: a tool for conservation. *Journal of Biogeography 30*(4), 607.

MacLean, S. J., and T. Seastedt (1979). Avian territoriality: Sufficient resources or interference competition? *American Naturalist 114*, 308–312.

Mangel, M., and C. W. Clark (1988). *Dynamic modeling in behavioral ecology*. Princeton, NJ: Princeton University Press.

Mangel, M., and B. D. Roitberg (1989). Dynamic information and host acceptance by a tephritid fruit fly. *Ecological Entomology 14*, 181–189.

Manly, B., L. Mcdonald, and D. Thomas (1993). *Resource Selection by Animals: Statistical Design and Analysis for Field Studies*. New York: Chapman and Hall.

Marsh, L., and R. Jones (1988). The form and consequences of random walk movement models. *Journal of Theoretical Biology 133*, 113–131.

Marzluff, J. M., S. T. Knick, and J. J. Millspaugh (2001). High-tech behavioral ecology: Modeling the distribution of animal activities to better understand wildlife space- use and resource selection. In J. Millspaugh and J. M. Marzluff (Eds.), *Radio tracking and animal populations*, pp. 310–326. San Diego: Academic Press.

Matthiopoulos, J. (2003). The use of space by animals as a function of accessibility and preference. *Ecological Modelling 159*, 239–268.

May, R. (1974). *Model Ecosystems*. Princeton, NJ: Princeton University Press.

Maynard-Smith, J. (1974). The theory of games and animal conflict. *Journal of Theoretical Biology 47*, 209–221.

Mech, L. D. (1991). *The way of the wolf*. Stillwater, MN: Voyageur Press.

Mech, L. D. (1966). *The wolves of Isle Royale*. U.S. Natl. Park Serv. Fauna Ser. 7.

Mech, L. D. (1977). Wolf pack buffer zones as prey reservoirs. *Science 198*, 320–321.

Mech, L. D. (1994). Buffer zones of territories of gray wolves as regions of interspecific strife. *Journal of Mammalogy 75*, 199–202.

Mech, L. D., and L. Boitani (2003). *Wolves: Behavior, ecology and conservation*. Chicago: University of Chicago Press.

Messier, F., and C. Barrette (1982). The social system of the coyote (*Canis latrans*) in forested habitat. *Canadian Journal of Zoology 60*, 1743–1753.

Mesterton-Gibbons, M., and E. S. Adams (2003). Landmarks in territory partitioning: A strategically stable convention? *The American Naturalist 161*, 685–697.

Mills, M. (1989). The comparative behavioral ecology of hyenas. In J. Gittleman (Ed.), *Carnivore behavior, ecology and evolution*, chapter 4. Ithaca, NY: Cornell University Press.

Mills, M., and M. Gorman (1987). The scent marking behavior of the spotted hyaena *Crocuta crocuta* in the southern Kalahari. *Journal of the Zoological Society of London 45*, 483–497.

Millspaugh, J. J., and J. M. Marzluff (2001). *Radio tracking and animal populations*. San Diego: Academic Press.

Moehlman, P. (1986). The ecology of cooperation in canids. In D. Rubenstein and R. Wrangham (Eds.), *Ecological aspects of social evolution*, pp. 64–86. Princeton, NJ: Princeton University Press.

Moehlman, P. (1989). Intraspecific variation in canid social systems. In J. Gittleman (Ed.), *Carnivore behavior, ecology and evolution*, pp. 143–163. Ithaca, NY: Comstock Publishing Associates.

Moorcroft, P. R. (1997). *Territoriality and carnivore home ranges*. Ph.D. thesis, Dept. of Ecology and Evolutionary Biology, Princeton University, Princeton, NJ.

Moorcroft, P. R., M. A. Lewis, and R. L. Crabtree (1999). Analysis of coyote home ranges using a mechanistic home range model. *Ecology 80*: 1656–1665.

Moorcroft, P. R., M. A. Lewis, and R. L. Crabtree (2006). Mechanistic Home Range Models Predict Spatial Patterns and Dynamics of Coyote Territories in Yellowstone. *Proceedings of the Royal Society Series B* (in press).

Murray, D., S. Boutin, and M. O'Donoghue (1994). Winter habitat selection by lynx and coyotes in relation to snowshoe hare abundance. *Canadian Journal of Zoology 72*, 1444–1451.

Murray, J. D. (1989). *Mathematical biology*. Biomathematics, vol. 19. Berlin: Springer-Verlag.

Murray, J. D. (2002). *Mathematical biology II: Spatial models and biomedical applications*. Biomathematics. New York: Springer-Verlag.

Myers, J., P. Connors, and F. Pitelka (1981). Optimal territory size and the sanderling. In A. Kamil and T. Sargent (Eds.), *Foraging behavior*. New York: Garland Press.

Nel, J. (1999). Social learning in canids: An ecological perspective. In H. O. Box and K. R. Gibson (Eds.), *Mammalian social learning: Comparative and ecological perspectives*. New York: Cambridge University Press.

Nelson, M., and L. D. Mech (1981). Deer organization and wolf predation in Minnesota. *Wildlife Monographs 77*, 1–53.

Nelson, M. E., and L. D. Mech (1986). Relationship between snow depth and gray wolf predation on white-tailed deer. *Journal of Wildlife Management 50*(3), 471–474.

Odum, E., and E. Kuenzler (1955). Measurement of territory and home range size in birds. *Auk 72*, 128–137.

Okarma, H., W. Jedrzejewski, K. Schmidt, S. Sniezko, A. Bunevich, and B. Jedrzejewska (1998). Home ranges of wolves in Bialowieza Primeval Forest, Poland, compared with other Eurasian populations. *Journal of Mammalogy 79*, 842–852.

Okubo, A. (1980). *Diffusion and ecological problems: Mathematical models*. Berlin: Springer-Verlag.

Okubo, A., and D. Grunbaum (2001). Mathematical treatment of biological diffusion. In *Diffusion and Ecological Problems: Modern Perspectives* (2d ed.), vol. 14 of *Interdisciplinary applied mathematics*. Berlin: Springer-Verlag.

Okubo, A., and S. Levin (2001). *Diffusion and ecological problems: Modern perspectives* (2d ed.), vol. 14 of *Interdisciplinary Applied Mathematics*. Berlin: Springer-Verlag.

Ostfeld, R. S. (1986). Territoriality and the mating system of California voles. *Journal of Animal Ecology 55*, 691–706.

Othmer, H., S. Dunbar, and W. Alt (1988). Models of dispersal in biological systems. *Journal of Mathematical Biology 26*, 263–298.

Othmer, H., and T. Hillen (2002). The diffusion limit of transport equations II: Chemotaxis equations. *SIAM Journal of Applied Mathematics 61*, 751–775.

Othmer, H., and A. Stevens (1997). Aggregation, blowup, and collapse: The ABC's of taxis in reinforced random walks. *SIAM Journal on Applied Mathematics 57*(4), 1044–1081.

Packard, J., and L. D. Mech (1980). Population regulation in wolves. In M. N. Cohen and H. Klein (Eds.), *Biosocial mechanisms of population regulation*, pp. 135–150. New Haven: Yale University Press.

Palsson, E., and H. Othmer (2000). A model for individual and collective cell movement in *Dictyostelium discoideum*. *PNAS 97*, 10448–10453.

Paquet, P. (1991). Scent marking behavior of sympatric wolves (*Canis lupus*) and coyotes (*Canis latrans*) in Riding Mountain National Park. *Canadian Journal of Zoology 69*(7), 1721–1727.

Paquet, P., and W. Fuller (1990). Scent marking and territoriality in wolves of Riding Mountain National Park. In R. Brown and D. W. Macdonald (Eds.), *Social odours in mammals*, vol. 4, chapter 10. Oxford: Clarendon Press.

Patlak, C. (1953). Random walk with persistence and external bias. *Bulletin of Mathematical Biophysics 15*, 311–338.

Person, D., and D. Hirth (1991). Home range and habitat of coyotes in a farm region of Vermont. *Journal of Wildlife Management 55*, 433–441.

Peters, R., and L. D. Mech (1975). Scent marking in wolves. *American Scientist 63*, 628–637.

Peterson, R., and R. Page (1988). The rise and fall of the Isle Royale wolves. *Journal of Mammalogy 69(1)*, 89–99.

Peterson, R., J. Woolington, and T. Bailey (1984). Wolves of the Kenai Peninsula, Alaska. *Wildlife Monographs 88*, 1–52.

Porter, W. F., and K. E. Church (1987). Effects of environmental patterns on habitat preference analysis. *Journal of Wildlife Management 51*, 681–685.

Post, E. (1999). Ecosystem consequences of wolf behavioral response to climate. *Nature 401*, 905–908.

Press, W., B. Flannery, S. A. Teukolsky, and W. Vetterling (1992a). *Numerical recipes in FORTRAN 77: The art of scientific computing*. Cambridge: Cambridge University Press.

Press, W., S. Teukolsky, W. Vetterling, and B. Flannery (1992b). *Numerical recipes in C: The art of scientific computing*. Cambridge: Cambridge University Press.

Priede, I., and S. Swift (1993). *Wildlife telemetry: Remote monitoring and tracking of animals*. New York: Ellis Horwood Ltd.

Ralls, K. (1971). Mammalian scent-marking. *Science 171*, 443–449.

Richardson, P. (1990). Scent marking and territoriality in the aardwolf. In R. Brown and D. W. Macdonald (Eds.), *Social Odours in mammals*, vol. 4, Chapter 8, pp. 378–387. Oxford: Oxford University Press.

Rodgers, A., R. Rempel, and K. Abraham (1996). A GPS-based telemetry system. *Wildlife Society Bulletin 24*, 559–566.

Rooney, S. M., A. Wolfe, and T. J. Hayden (1998). Autocorrelated data in telemetry studies: Time to independence and the problem of behavioural effects. *Mammal Review 28*, 89–98.

Rothman, J. and L. D. Mech (1979). Scent marking in lone wolves and newly formed pairs. *Animal Behavior 27*, 750–760.

Roughgarden, J. (1995). *Anolis lizards of the Carribbean: Ecology, evolution and plate tectonics*. Oxford: Oxford University Press.

Rubenstein, D., and R. Wrangham (1986). *Ecological aspects of social evolution*. Princeton, NJ: Princeton University Press.

Saad, Y., and M. Schultz (1986). GMRES: A generalized minimal residual algorithm for solving nonsymmetric linear systems. *SIAM Journal on Scientific Computing 7*, 856–869.

Samuel, M., D. Pierce, and E. Garton (1985). Identifying areas of concentrated use within the home range. *Journal of Animal Ecology 54*, 711–719.

Schaller, G. (1972). *The Serengeti Lion: A study of predator-prey relations*. Chicago: University of Chicago Press.

Schiesser, W. (1991). *The method of lines*. New York: Academic Press.

Schoener, T. W. (1981). An empirically based estimate of home range. *Theoretical Population Biology 20*, 281–325.

Schoener, T. W. (1983). Simple models of optimal feeding-territory size: A reconciliation. *American Naturalist 121(5)*, 608–629.

Schoener, T. W., and A. Schoener (1980). Density, sex ratio and population structure in some Bahamian *anolis* lizards. *Journal of Animal Ecology 49*, 19–53.

Segel, L. (1972). Simplification and scaling. *SIAM Review 14*, 547–571.

Sheldon, J. (1992). *Wild dogs: Natural history of the non-domestic canidae*. San Diego: Academic Press.

Shenbrot, G., B. Krasnov, and K. Rogovin (1999). *Spatial Ecology of Desert Rodent Communities*. Berlin: Springer-Verlag.

Siniff, D., and C. Jessen (1969). A simulation model of animal movement patterns. *Advances in Ecological Research 6*, 185–219.

Skellam, J. (1951). Random dispersal in theoretical populations. *Biometrika 38*, 196–218.

Skellam, J. (1973). The formulation and interpretation of mathematical models of diffusionary processes in population biology. In M. Bartlett and R. Hiorns (Eds.), *The mathematical theory of the dynamics of biological populations*. New York: Academic Press.

Smith, D., R. O. Peterson, and D. B. Houston (2003). Yellowstone after wolves. *BioScience 53*, 330–340.

Smoluchowski, M. (1916). Drei Vorträge über Diffusion, brownsche Bewegung und Koagulation von Kolloidteilchen. *Physikalische Zeitschrift 17*, 557–585.

Stamps, J. (1977). Social behavior and spacing patterns in lizards. In C. Gans and D. Tinkle (Eds.), *Biology of the reptilia* (1st ed.), vol 7: Ecology and Behavior. New York: Academic Press.

Stamps, J., M. Beuchner, and V. Krishnan (1987). The effects of habitat geometry on territorial defense costs: Intruder pressure in bounded populations. *American Zoologist 27*, 307–325.

Swihart, R., and N. Slade (1985). Testing for the independence of observations in animal movements. *Ecology 66*, 1176–1184.

Swihart, R., and N. Slade (1997). On testing for independence of animal movements. *Journal of Agricultural, Biological, and Environmental Statistics 2*, 1–16.

Taylor, R. J., and P. J. Pekins (1991). Territory boundary avoidance as a stabilizing factor in wolf-deer interactions. *Theoretical Population Biology 39*, 115–128.

Thomas, D. L., and Taylor, J. (1990). Study designs and tests for comparing resource use and availability. *J. Wildl. Manage.* 54, 322–330.

Tranquillo, R., and W. Alt (1990). Glossary of terms concerning oriented movement. In W. Alt and G. Hoffmann (Eds.), *Biological motion*, no. 89 in *Lecture Notes in Biomathematics*, pp. 510–517. Berlin: Springer-Verlag.

Turchin, P. (1991). Translating foraging movements in heterogeneous environments into the spatial distribution of foragers. *Ecology 72*, 1253–1266.

Turchin, P. (1998). *Quantitative analysis of movement: Measuring and modeling population redistribution in animals and plants*. Sunderland, MA: Sinauer.

Van Ballenberghe, V. (1972). *Ecology, movements and population characteristics of timber wolves in northeastern Minnesota*. Ph.D. thesis, University of Minnesota.

Van Ballenberghe, V., A. W. Erickson, and D. Byman (1975). Ecology of the timber wolf in northeastern Minnesota. *Wildlife Monographs 43*, 1–43.

Waser, P. M. (1985). Spatial structure in mangabey groups. *International Journal of Primatology 6*, 569–580.

Wells, M., and M. Bekoff (1981). An observational study of scent marking in coyotes. *Animal Behavior 29*, 332–350.

White, G., and R. Garrott (1997). *Analysis of wildlife radio tracking data*. Boston: Academic Press.

White, K. A. J., M. A. Lewis, and J. D. Murray (1996). A model for wolf-pack territory formation and maintenance. *Journal of Theoretical Biology 178*, 29–43.

White, K. A. J., M. A. Lewis, and J. D. Murray (1998). On wolf territoriality and deer survival. In J. Bascompte and R. V. Solé (Eds.), *Modeling spatiotemporal dynamics in ecology*, chapter 6, pp. 105–126. Berlin: Springer-Verlag.

White, K. A. J., J. D. Murray, and M. A. Lewis (1996). Wolf-deer interactions: A mathematical model. *Proc. Roy. Soc. Lond. B 263*, 299–305.

Woodroffe, R. (2001). Strategies for carnivore conservation: Lessons from contemporary extinctions. In J. L. Gittleman, S. M. Funk, D. Macdonald, and R. K. Wayne (Eds.), *Conservation biology 5: Carnivore conservation*, pp. 61–92. Cambridge: Cambridge University Press.

Woodroffe, R., and J. Ginsberg (1998). Edge effects and the extinction of populations inside protected areas. *Science 280*, 2126–2128.

Worton, B. (1987). A review of models of home range for animal movement. *Ecological Modeling 38*, 277–298.

Worton, B. (1989). Kernel methods for estimating the utilization distribution in home range studies. *Ecology 70(1)*, 164–168.

Zollner, P., and S. Lima (1999). Search strategies for landscape-level inter-patch movements. *Ecology 80*, 1019–1030.

Index

advection, relationship to directed motion, 13–15

advection coefficient, in advection-diffusion equation with directional persistence, 141; in the conspecific avoidance model, 41; in generalized advection-diffusion equations, 10–12, 145–146; in the prey availability plus conspecific avoidance model, 87; in the steep terrain avoidance plus conspecific avoidance model, 81

advection-diffusion equation(s), 7, 21–22; derivation from stochastic movement process, 137–138; with directional persistence, 140–141

African wild dogs, 132

aggression, role in determining evolutionarily stable movement strategy, 126–128

allopatric patterns of space use in carnivores, 102–103

badger, scent mark patterns, 63, 65

bias parameter, b, in the conspecific avoidance model, 40

biased random walk model. *See* localizing tendency model

bivariate normal distribution. *See* statistical home range models

borderland marking. *See* scent marks, spatial distribution of

boundary conditions, 21; absorbing, in the localizing tendency model, 153–156; for the conspecific avoidance model, 44; effects on patterns of space use predicted at Hanford, 47

buffer zones, 97–100, 102; conditions for the formation of, 72–78, 134

carnivores, community in Yellowstone National Park, 132; conservation and management, 131

circular normal distribution. *See* Von Mises distribution

conspecifics, effects on coyote patterns of space use at Hanford, 37

conspecific avoidance model, analysis in 1D, 67–78; β parameter in, 43, 58–59; biological motivation, 38–39; bounded avoidance response, 74–75; bounded scent marking response, 75–77; derivation from stochastic movement process, 144–148; differences in 1D and 2D, 65–66; effects of population density, 61–64; empirical evaluation at Hanford, 43–48, 52–54; equations for space use in 2D, 41–44; influence of neighboring packs, 56; model formulation, 39–43; model predictions at Hanford, 48–53, 133–134; numerical simulations, 45; pattern of space use in absence of overmarking, 68–69, 77–78; predicted effects of pack removal at Hanford, 50–52; predicted patterns of fine-scale movement at Hanford, 49–51

correlated random walk, 4

correlation, between successive movement directions, 16–17, 27–28, 140–141; between successive relocations, 24, 16–17, 105, 153–156

coyotes, foraging behavior, 84–85; home range patterns at Hanford, 35–37, 43–54, 133; home range patterns in Yellowstone National Park, 79–91, 133–134; home range size, 1–2, 23; interactions with wolves in Yellowstone National Park, 132; scent mark patterns, 60–64; temporal variation in prey availability. *See also* temporal variability, 130

den site, absence of, 92

diffusion, isotropic assumption, 13–15; relationship to random motion, 15–22

diffusion coefficient, in advection-diffusion equation with directional persistence, 141; in the conspecific avoidance model, 41; in generalized advection-diffusion equations, 10–12, 146; in home range model without a den site, 93; in the localizing tendency model, 30; in the prey availability plus conspecific avoidance model, 87; in the

MONOGRAPHS IN POPULATION BIOLOGY
EDITED BY SIMON A. LEVIN AND HENRY S. HORN

MONOGRAPHS IN POPULATION BIOLOGY

EDITED BY SIMON A. LEVIN AND HENRY S. HORN

(continued)

Milton Keynes UK
Ingram Content Group UK Ltd.
UKHW021149140824
446917UK00001B/2